Praise for *A Growers Guide for Balancing Media-Based Soils*

At Tap Root Fields we are dedicated to farming in a way that honors our local ecology, and we strive to improve our soils for the next generation. The land on which TRF set out to establish our mixed vegetable farm had been deemed by local farmers as degraded and only good for hay. In the last few years, we have done regular soil testing and implemented the organic-certified soil amendment recommendations as advised by Bill McKibben. The year we started this practice, we saw dramatic improvements in crop yields and disease resistance.
Ben Zoba
Tap Root Fields

The fact is that most people don't know how to interpret soil tests. Bill McKibben increased my understanding of the chemical/nutrient portion of soil tenfold. These tests help as an indicator of overall plant nutritional health and serve as an excellent basis for diagnosing whether existing problems are nutritional in nature. Bill was pivotal in assisting us in the development of our current fertigation process and nutrient composition in our living soil. Grow Ohio's priority from the beginning was locking in a process for our living-soil product that was consistent and repeatable. Bill helped us achieve this. We were able to increase total terpenes by over 31.5 percent as well as total cannabinoids by 23.7 percent compared to our traditional coco-grown rounds.
Edward T. Gladstone
Director of R/D, Grow Ohio

Successful indoor cultivation depends on building a living system where microbial diversity drives nutrient cycling, plant resilience, and overall balance. Having an agronomist who shares this vision has been essential to our work, and Bill McKibben's mineral-balancing approach has proven to be the most effective framework we've used. His method combines simplicity with scientific rigor, optimizing cation ratios and harmonizing nutrient availability in ways that create a healthier rhizosphere for beneficial microorganisms. Over the years, we have tested many balancing strategies, but Bill's stands out for its consistency and measurable results: stronger plants, greater nutrient uptake efficiency, and improved resistance to stress. His approach doesn't just complement our microbial strategies—it amplifies them, giving us both practical reliability and long-term sustainability in our indoor living systems.

DJ Mumm
Cultivation Direction, Sensicare Dispensary

We used to have lots of crop loss. People told us we were missing potassium or phosphorus or nitrogen, and it was very confusing. But in reality these nutrients were there—they just weren't available for the plant to take up. Adding more and more just hurt the crops. Bill McKibben saw that my salt level was too high and that the soil media density was too compressed. He helped us break it down by adding peat moss and giving it plenty of aeration and balancing everything so the plant was able to take up the nutrients. Since then we've had better results. We appreciate all of Bill's help and knowledge!

Beatrice Aviles
Cannabis grower

Growing cannabis in the Northeast is challenging: the season is short, and every week matters. To make the most of that narrow window, we must focus on establishing vigorous growth from the very start. Our approach begins with comprehensive testing of both soil contents and nutrient solubility. We use this data to determine what is needed to balance the nutrients in the soil and make sure they are readily available to our plants. The goal is to create a soil environment where plants never struggle for essentials. By balancing the soil to targeted levels, we eliminate limiting growth factors and ensure that crops don't expend unnecessary energy on resource acquisition and instead channel it directly into growth and vigorous production. Careful soil balancing over the years, with the help of Bill McKibben, has allowed us to increase productivity, significantly reduce input costs, and improve soil health year over year. I know without a doubt that without rigorous testing to begin and end each season it would not be possible to maintain a balanced soil profile and deliver consistent quality and yields despite the many constraints and tough conditions we face here.

Jesse Sgambati
Director of operations, EOS Farms

Stephen Filipiak works media-based soil for turmeric production in Hawaii.
(Courtesy of Stephen Filipiak)

A Growers Guide
for
BALANCING MEDIA-BASED SOILS

A Practical Guide to Interpreting Tests
of Modified Growing Media

William "Crop Doc" McKibben

Foreword by Susan Shaner

About Acres U.S.A.

Founded in 1971 by Charles Walters, *Acres U.S.A.* magazine emerged from the need to promote ecological farming practices in a time when industrial agriculture was heavily reliant on synthetic fertilizers and pesticides. Inspired by figures like Rachel Carson and Dr. William Albrecht, Walters used the magazine, and later books and conferences, to advocate for sustainable agriculture that prioritized soil health and natural processes. Acres U.S.A. provided a platform for these ideas and helped to popularize alternative methods like cover cropping and integrated livestock management.

Though the agricultual landscape still relies heavily on conventional methods, Acres U.S.A. has been instrumental in the growing movement toward regenerative agriculture. By disseminating knowledge and supporting eco-conscious farmers, the company continues to champion sustainable practices through its publications, conferences, and online resources, contributing to a shift toward a more ecologically sound approach to farming.

Find Out More About Acres U.S.A.

Subscribe to the Online or Print Magazine
https://go.acresusa.com/magazine39971

Attend Our Eco-Ag Conference
https://go.acresusa.com/events39971

Visit the Acres U.S.A. Bookstore
https://go.acresusa.com/bookstore39971

Join the Free Newsletter
https://go.acresusa.com/newsletters39971

A Growers Guide for Balancing Media-Based Soils

A Practical Guide to Interpreting Tests of Modified Growing Media

Copyright © 2025 by William L. McKibben

All rights reserved. No part of this book may be used or reproduced without written permission except in cases of brief quotations embodied in articles and books.

The information in this book is true and complete to the best of our knowledge. All recommendations are made without guarantee on the part of the author and Acres U.S.A. The author and publisher disclaim any liability in connection with the use or misuse of this information.

PO Box 351
Viroqua, WI 54665 U.S.A.
512-892-4400 • info@acresusa.com • www.acresusa.com

Printed in the United States of America

ISBN: 978-1-60173-997-1

Dedication

I dedicate this book to my Dad, a World War II veteran, who died July 25, 2018, and who helped me in my business many a day. Thanks, Dad.

Table of Contents

Acknowledgements . . . . *xv*

Foreword . . . . *xvii*

Introduction . . . . *xix*

CHAPTER 1:
Making Your Own Growing Media 1

CHAPTER 2:
Analysis Needed for a Successful Growing Medium . 19

CHAPTER 3:
The Impact of Nutrients on Plant Growth . . . 37

CHAPTER 4:
Water and Its Effect on Growing Media 57

CHAPTER 5:
Nutrient Balancing in Media 63

CHAPTER 6:
Rebalancing a Problem Mix with Peat 83

Conclusion 97

Index 99

Acknowledgements

This book would not be possible without all the wonderful people that I have had the privilege of working with over the past forty-nine years. Over the past ten years, working with media growers has truly been a learning experience, and I thank them for the opportunity to learn and grow with them. I quickly learned that growing in media is not as easy as it first appears. This book is a summation of the knowledge I have gleaned from the people I have worked with.

I would like to thank the people from Logan Labs who have supported me in writing this book, as well as my previous books, with all sorts of lab analysis. I would especially like to thank Susan Wallace at the lab, who helped to reprint many of the examples used in my books.

I would also like to thank the people at Mulch Makers of Ohio for allowing me to use some of their test results, as well as helping me put together some bulking agents and mixes for testing at various grow facilities. Their cooperation and attention to detail has been nothing short of amazing. Thank you, Keith and Matt.

Foreword

It has been my pleasure to know Bill McKibben, "The Crop Doc," since 1984. We met when I began working in a soil laboratory. He was one of the first independent soil agronomists I encountered.

Bill has always been very inquisitive. He is never content with the status quo. He is always looking for a better way to determine soil fertility and to improve yields. He was instrumental in incorporating the use of saturated paste analysis in his business and also encouraging others to do the same. In 2002, I opened my own independent lab. Bill has been a great asset, providing data interpretation and recommendations to our clients.

In my first thirty-five years of experience in the agriculture industry, nothing new came along other than some improvements in testing equipment and some new testing methods. Crops and soils were the same. With the arrival of the cannabis industry, though, along came some new mediums that we soon learned were not the same as the soils we had been testing for decades. We were now working with manmade media. Several terms are used for these mediums, such as soilless media, living soil, organic soil, etc.

Bill directed the research that was necessary to determine the best testing procedures for these media. After several heated debates with Bill, we came to a consensus on the best way to test them. Bill determined that the weight of the medium must be taken into consideration to make the correct recommendation for amendments.

Bill's dedication to never being happy with the status quo shows in the success of his clients.

This book is a must read for the living soil, organic soil, and soilless medium community.

Susan Shaner
Director (retired) of Logan Labs

Introduction

Soil amended to above 12 percent organic matter needs to be balanced differently than normal field soils.
(Courtesy of Tap Root Fields, Ben Zoba)

For nearly fifty years, I have been consulting on crops grown in all types of natural soils. This myriad of crops—from commercial corn, beans, potatoes, and tomatoes to pecans, coffee, dahlias, and roses—have mostly been grown in native soils in various climates, using a variety of cultural practices. Seldom did someone who came to me have a total cropping failure. Generally, they had a problem with overall production—whether it was the result of compaction on the soil physics side or nutrient deficiencies resulting in poor production and increased insect and disease pressure. Sometimes seed germination was poor, or plants started slowly and couldn't handle various stressful situations.

After getting a soil test, I used various methods to improve the soil nutrient balance, based on the Albrecht philosophy on high cation exchange capacity (CEC) soils or the Strategic Level of Available Nutrients (SLAN) approach on low CEC soils. Cation exchange capacity, also known as total exchange capacity (TEC), is a measure of the soil's capacity to hold and exchange positively charged plant nutrients like calcium, magnesium, and potassium. A high CEC soil has a cation exchange capacity greater than 10 milliequivalents of negative charge per 100 grams of oven-dried soil (meq/100 g), and a low CEC soil has a CEC less than 10 meq/100 g. Soils with a CEC between 7–12 meq/100 g were best served by using a combination of both balancing strategies.

I measured my success rate by the number of returning clients. I worked with many of my local clients for over thirty years. The goal was to achieve new highs in yield and quality and minimize the lows in stressful years. Initially, this was accomplished with just a standard soil test from Logan Labs in Ohio, but as yields got better, more information was needed. That was why I began including paste analysis and tissue analysis in the soil test package.

Over the past ten years, primarily due to the cannabis industry, I have worked with a large number of growers using growing media, which is also known as "growing substrate," "potting mix," or "soilless media." Many growers like to refer to their mix as a "living soil," but virtually all crops grown in either a natural soil or a growing medium, with maybe the exception of hydroponics, are grown in a substrate containing living organisms. Living organisms are extremely important in crop production and contribute what I like to call the X-factor to yield and—most importantly—quality. Just compare a field-grown tomato and a hothouse tomato, and you'll know what I mean.

What exactly am I referring to when I talk about growing media? There are two types of growing media. The first is a soilless mix of various organic components, like peat and an assortment of composts. There is virtually no natural soil in the mix. The second is a natural soil that has been modified by the addition of various organic materials, such as peat and composts of various products such as leaves, bark, manure, etc. These materials have been added to such a level that they significantly change the bulk density. On the Logan Labs report, the density of growing media is referred to as media density. When enough organic material has been added to a soil to reduce the media density from 1.0 g/cm^3, which is within the range for a normal soil, to a level of less than 0.75 g/cm^3, I consider it to be a growing medium. The standard test for soil organic matter will usually be over 12–15 percent for this type of soil. The non-soil mixes will almost always be over 20 percent organic matter.

An example of the type of growing media discussed in this book — from a cannabis cultivator in Ohio.
(Courtesy of Grow Ohio)

Whatever you choose to call your growing media is fine, but in this book I will be talking about soil chemistry because I am a soil chemist, not a microbiologist. My goal is to create a soil chemistry environment that is very conducive to microbial life. Microbial activity plays many significant roles in increasing nutrient availability. For example, microbes are essential for increasing phosphorus availability and breaking down elemental sulfur to lower pH.

I mentioned earlier that farmers growing in natural soils rarely have a total crop failure—barring a drought, flood, or some act of God—but it seemed as though growers using growing media with controlled watering, light, and temperature were having a higher percentage of crop failures. After looking at a number of test results, I quickly realized that growers were trying to treat their growing media like a natural soil—and that has proved to be WRONG.

A quick example, which will be discussed in detail later, is the matter of bulk density. After looking at a number of soil test results with extremely high levels of nutrients and consequently high salts, it occurred to me that people were not taking into account the bulk density of their mix and were overfertilizing. After discussing this issue with Susan Shaner, director of Logan Labs, we decided to include media density on all standard soil tests properly noted on the worksheet as growing media.

In this book I discuss some of the dos and don'ts of making your own mix. There are an unlimited number of ways to put together a growing medium, but hopefully I can point out some of the pitfalls of putting a mix together and increase your chance of success. I show a number of test results from ingredients that I have tested, as well as a number of final products on the market, many of which have different problems. There is no guarantee that buying a mix from a big box store or greenhouse will be the safest and easiest path to a successful grow. I show you the best test methods for identifying good or bad mixes. I give you my suggestions for the best nutrient levels needed to produce a successful crop in growing media. I admit that my desired levels are more conservative than many of the growing media producers in the marketplace. I also discuss the individual soil nutrients and their role in crop production. Lastly, I discuss irrigation water—whether it be reverse osmosis, well, or municipal water—and its effect on nutrient movement and buildup over the growing cycle.

This is a lot to cover, so let's get started.

Chapter 1:

Making Your Own Growing Media

A well-balanced growing bed after harvest.
(Courtesy of Grow Ohio)

Making your own growing media is not as easy as one might think. The only thing that's harder is trying to duplicate the mix six months or a year later. I think this will become clear as we go through this section.

I like to divide the ingredients for a mix into three main categories: major bulking agents, minor bulking agents, and mineral supplements.

The major bulking agents include compost, peat, and mulches. Mulches include ground hardwood or pine bark and leaf mulch. These materials should only be used by a large media-producing company since they are raw and not composted yet. Raw ingredients such as ground bark fines have a wide carbon:nitrogen ratio and will tie up nitrogen in a growing medium. An individual grower should not use these products unless they are already composted. It is unlikely that a small grower can accumulate enough of the raw products to create a big enough pile to generate the heat necessary to completely compost them. No matter what source of material is used, raw ingredients need to be completely composted before using in a growing medium.

If you are using composted products, such as leaves or ground bark, be sure to get around 25 percent extra beyond what you need for your beds or containers. The extra can be used later if your beds or containers need to be diluted due to an accumulation of salts from irrigation water or too many additives during the various growing cycles. This will be discussed further in the water section.

You might ask, why not save money by only buying as much compost as you need now? If you need more later, can't you just go back to the same place where you originally bought it and get some more? I'm glad you asked.

When buying composts—whether they be plant-based, manure-based, or some combination—it is hard to be sure that the ingredients or ratio of ingredients in a compost from any one source hasn't changed. In the fall, a compost of leaves and bark may have a higher percentage of leaves to bark compared to spring or late summer when leaves may be running out.

A manure-based compost from chickens should be fairly consistent since the feeding programs are highly regulated and seldom changed; however, compost from egg-laying chicken manure will be different from compost made from the manure of meat chickens or pullets. Cattle may shift from a high-legume diet to one higher in corn silage, depending on the weather and its effect on the hay and corn crops. Manure from lactating dairy cows will generally be richer in nutrients than manure from dry cows or heifers, due to the higher level of concentrates in the milking ration. Beef cattle manure will generally have the highest concentration of nutrients, since the ration contains a higher level of grain.

Figures 1A and 1B show the soil test differences between two dairy manure composts. The level of nutrients are nearly four times higher in the old manure than the new manure.

The old manure could easily result in a high level of soluble salts, depending on the percentage used in the growing mix. Figures 2A and 2B show the test results when different percentages of this old manure compost was used in a growing medium. Blend 1 used 10 percent manure compost and Blend 2 used 20 percent. It turned out that 15 percent manure was the sweet spot for this growing mix.

When starting off making a bulking mix, the K.I.S.S. method is probably the best idea: "Keep It Simple, Stupid." Locate three or four easily-acquired bulking products such as peat, manure compost, and plant-based compost. This will make it easier to repeat the mix later on. It would be nice to have analytical data on each of these products, but that is usually not available and even if it is, you need to inquire how recently the material has been tested. A company that tests on a regular basis will be more concerned with consistency. Look at the last two or three lab results and see if there is good consistency.

A manure-based compost will supply most of the major nutrients—nitrogen, phosphorus, and potassium. One of the biggest problems that I see is the use of too much manure compost in a mix, which results in nutrient imbalances and high salt concentrations. A rule of thumb is to use manure compost at a maximum rate of 10–20 percent by volume. Chicken or dairy compost with minimal bedding (straw or sawdust) should probably be limited to 10–15 percent. Start off with

An example of plants experiencing stress from high salt concentrations. (Courtesy of Randy Shami)

Sample Location			Old Manure Mulch	New Manure
Sample ID				
Lab Number			1	2
Sample Depth in inches			9	6
Total Exchange Capacity (M. E.)			16.83	7.10
pH of Soil Sample			9.2	10.0
Organic Matter, Percent			>20	>20
ANIONS	SULFUR:	p.p.m.	51	38
	Mehlich III Phosphorous:	as (P_2O_5) lbs / acre	1242	292
EXCHANGEABLE CATIONS	CALCIUM: lbs / acre	Desired Value		
		Value Found	2999	910
		Deficit		
	MAGNESIUM: lbs / acre	Desired Value		
		Value Found	1622	470
		Deficit		
	POTASSIUM: lbs / acre	Desired Value		
		Value Found	6534	1727
		Deficit		
	SODIUM:	lbs / acre	945	256
BASE SATURATION %	Calcium (60 to 70%)		29.708	32.04
	Magnesium (10 to 20%)		26.77	27.58
	Potassium (2 to 5%)		33.19	31.19
	Sodium (.5 to 3%)		8.14	7.83
	Other Bases (Variable)		2.20	1.40
	Exchangable Hydrogen (10 to 15%)		0.00	0.00
TRACE ELEMENTS	Boron (p.p.m.)		1.41	1.54
	Iron (p.p.m.)		55	80
	Manganese (p.p.m.)		10	7
	Copper (p.p.m.)		1.43	0.68
	Zinc (p.p.m.)		13.6	2.76
	Aluminum (p.p.m.)		8	11
OTHER	Cobalt ppm		0.038	
	Molybdenum ppm		< 0.02	
	Ammonium (p.p.m.)		1.3	0.2
	Nitrate (p.p.m.)		116.9	3.4
	Selenium ppm		0.26	
	Silicon ppm		8.2	
	Estimated Nitrogen Release #N/Acre		N/A	
	EC mmhos/cm		2.03	
	Media Density g/cm3			0.14

Figure 1A: *Standard soil test results for two dairy manure composts*

Sample Location			Old Manure Mulch	New Manure
Sample ID				
Lab Number			207848	224046
Water Used			DI	DI
pH			9.2	10.0
Soluble Salts		ppm	726	783
Chloride (Cl)		ppm	326	372
Bicarbonate (HCO3)		ppm	360	489
ANIONS	SULFUR	ppm	7.3	44.4
	PHOSPHORUS	ppm	5.76	11.29
SOLUBLE CATIONS	CALCIUM	ppm	9.88	8.68
		meq/l	0.49	0.43
	MAGNESIUM	ppm	15.59	35.59
		meq/l	1.30	2.97
	POTASSIUM:	ppm	273.80	153.00
		meq/l	7.11	3.97
	SODIUM	ppm	56.18	112.00
		meq/l	2.44	4.87
PERCENT	Calcium		4.35	3.54
	Magnesium		11.45	24.22
	Potassium		62.67	32.46
	Sodium		21.53	39.77
TRACE ELEMENTS	Boron (p.p.m.)		0.14	0.5
	Iron (p.p.m.)		1.1	6.46
	Manganese (p.p.m.)		0.06	0.18
	Copper (p.p.m.)		0.25	0.18
	Zinc (p.p.m.)		0.05	0.12
	Aluminum (p.p.m.)		0.07	0.23
OTHER				

Figure 1B: *Paste test results for two dairy manure composts*

Sample Location			Blend	Blend
Sample ID			1	2
Lab Number			30	31
Sample Depth in inches			9	9
Total Exchange Capacity (M. E.)			7.90	14.70
pH of Soil Sample			6.8	6.9
Organic Matter, Percent			>20	>20
ANIONS	SULFUR:	p.p.m.	118	220
	Mehlich III Phosphorous:	as (P_2O_5) lbs / acre	156	407
EXCHANGEABLE CATIONS	CALCIUM: lbs / acre	Desired Value	3223	5995
		Value Found	3090	5472
		Deficit	-133	-523
	MAGNESIUM: lbs / acre	Desired Value	341	634
		Value Found	417	907
		Deficit		
	POTASSIUM: lbs / acre	Desired Value	369	687
		Value Found	815	1989
		Deficit		
	SODIUM:	lbs / acre	204	328
BASE SATURATION %	Calcium (60 to 70%)		65.19	62.06
	Magnesium (10 to 20%)		14.66	17.14
	Potassium (2 to 5%)		8.82	11.57
	Sodium (.5 to 3%)		3.73	3.24
	Other Bases (Variable)		4.60	4.50
	Exchangable Hydrogen (10 to 15%)		3.00	1.50
TRACE ELEMENTS	Boron (p.p.m.)		0.76	1.3
	Iron (p.p.m.)		93	162
	Manganese (p.p.m.)		13	21
	Copper (p.p.m.)		0.25	0.47
	Zinc (p.p.m.)		2.7	5.54
	Aluminum (p.p.m.)		39	55
OTHER	Cobalt ppm		0.079	0.125
	Molybdenum ppm		< 0.02	< 0.02
	Ammonium (p.p.m.)		0.7	0.5
	Nitrate (p.p.m.)		11.6	27.7
	Selenium ppm		< 0.02	0.16
	Silicon ppm		4.9	6.2
	Estimated Nitrogen Release #N/Acre		N/A	N/A
	EC mmhos/cm		0.49	0.66

Figure 2A: *Standard soil test results for different percentages of old manure compost in a growing medium*

Sample Location			Old Manure Mulch	New Manure
Sample ID				
Lab Number			207848	224046
Water Used			DI	DI
pH			9.2	10.0
Soluble Salts		ppm	726	783
Chloride (Cl)		ppm	326	372
Bicarbonate (HCO3)		ppm	360	489
ANIONS	SULFUR	ppm	7.3	44.4
	PHOSPHORUS	ppm	5.76	11.29
SOLUBLE CATIONS	CALCIUM	ppm	9.88	8.68
		meq/l	0.49	0.43
	MAGNESIUM	ppm	15.59	35.59
		meq/l	1.30	2.97
	POTASSIUM:	ppm	273.80	153.00
		meq/l	7.11	3.97
	SODIUM	ppm	56.18	112.00
		meq/l	2.44	4.87
PERCENT	Calcium		4.35	3.54
	Magnesium		11.45	24.22
	Potassium		62.67	32.46
	Sodium		21.53	39.77
TRACE ELEMENTS	Boron (p.p.m.)		0.14	0.5
	Iron (p.p.m.)		1.1	6.46
	Manganese (p.p.m.)		0.06	0.18
	Copper (p.p.m.)		0.25	0.18
	Zinc (p.p.m.)		0.05	0.12
	Aluminum (p.p.m.)		0.07	0.23
OTHER				

Figure 2B: *Paste test results for different percentages of old manure compost in a growing medium*

2.7–4.0 cubic feet of compost per cubic yard of mix. Figures 2A and 2B show the difference 10 percent manure can make in a blend of plant-based compost, peat, and manure. There is a substantial increase in soluble salts—160 ppm—primarily due to the increase in potassium and sulfur. Fifteen percent resulted in the best combination.

The third ingredient in our bulking mix is peat. There are many grades of peat, based on the length of aging and formative plant species. Peat is formed under anaerobic (lack of oxygen) conditions and will develop over time into a highly organic (greater than 20 percent organic matter) and acidic composite of decomposed plants. Peat contributes primarily to the mix by holding water, providing aeration, and serving as a rooting medium. Sphagnum peat moss is formed from sphagnum mosses, while most other peats are formed from plants like cattails, sedges, and various grasses. Peat will add a lot of organic matter to the mix, but really does not add anything to the nutrient-holding capacity and will not release nitrogen. Soil testing labs often give you an estimated nitrogen release (ENR) based on organic matter levels. These nitrogen numbers cannot be counted on in a growing medium. I prefer a peat that is black—the blacker the better. This may be just personal preference, but the blacker peats don't seem to be as hydrophobic when they dry out. If high-peat growing media do dry out, rewetting them can be made easier by using a soil surfactant.

Earlier, I discussed inconsistencies with different compost products, and peat samples are no different. Figure 3 shows the test results from two peat samples from the same bog, one at a 4-foot depth and another at a 20-foot depth. The 4-foot sample may be a couple of hundred years old, while the 20-foot sample could be more than a thousand years old.

Bogs are generally in low-lying areas that are prone to flooding, and this bog is no different. The shallow peat sample has a lot higher calcium level than the deep sample. This is probably due to farming practices over the last hundred years. Calcium in the soil is mass-flow mobile and easily lost through erosion and leaching, which can be caused by tillage and fertilizer applications of nitrogen and potassium. Sulfur levels are generally high in soils that are saturated throughout the year. Under saturated conditions, nitrogen will denitrify and gas off, but sulfur tends

Sample Location			#2 Peat	Peat
Sample ID			4'	20'
Lab Number			26	27
Sample Depth in inches			9	6
Total Exchange Capacity (M. E.)			53.73	35.43
pH of Soil Sample			5.3	5.1
Organic Matter, Percent			>20	>20
ANIONS	SULFUR:	p.p.m.	1454	558
	Mehlich III Phosphorous:	as (P_2O_5) lbs / acre	36	24
EXCHANGEABLE CATIONS	CALCIUM: lbs / acre	Desired Value		
		Value Found	15714	6230
		Deficit		
	MAGNESIUM: lbs / acre	Desired Value		
		Value Found	1433	524
		Deficit		
	POTASSIUM: lbs / acre	Desired Value		
		Value Found	72	36
		Deficit		
	SODIUM:	lbs / acre	347	88
BASE SATURATION %	Calcium (60 to 70%)		48.74	43.96
	Magnesium (10 to 20%)		7.41	6.16
	Potassium (2 to 5%)		0.11	0.13
	Sodium (.5 to 3%)		0.94	0.53
	Other Bases (Variable)		6.80	7.20
	Exchangable Hydrogen (10 to 15%)		36.00	42.00
TRACE ELEMENTS	Boron (p.p.m.)		1.62	0.99
	Iron (p.p.m.)		414	473
	Manganese (p.p.m.)		29	17
	Copper (p.p.m.)		< 0.2	0.53
	Zinc (p.p.m.)		2.94	2.33
	Aluminum (p.p.m.)		218	192
OTHER	Cobalt ppm		0.322	0.154
	Molybdenum ppm		< 0.02	0.1
	Ammonium (p.p.m.)		3.4	0.2
	Nitrate (p.p.m.)		5.9	1
	Selenium ppm		0.04	0.64
	Silicon ppm		13.8	6
	Carbon %			17.23
	C/N			14.59
	Nitrogen			11800

Figure 3: *Standard soil test results for two peat samples*

to accumulate. When looking at a standard soil report that has a high level of sulfur and no recent sulfur application, poor drainage is my first concern.

Mulch, compost, and peat make a lightweight mix, oftentimes 600–700 pounds per cubic yard, three or four times lighter than a cubic yard of soil. If you desire a heavier mix, you can add topsoil, sand, or pumice. Topsoil could have weed seeds, pesticides, or disease organisms coming along for the ride. If you are growing organically, I would consider staying away from adding soil to your mix, even if it comes from a certified organic farm. Sand is a better choice, but be sure to use silica sand and not calcareous sand. Calcareous sands will add too much calcium, resulting in a high pH. The calcium in the sand will continually break down, making soil balancing extremely difficult. A silica sand will primarily contribute weight, not nutrients.

Figure 4A and 4B shows four sands. The foundry sand is the only silica sand of the group, as you can see from the low total exchange capacity (TEC) and the low calcium value. The pH levels are all high, but the impact on a mix from the foundry sand will be negligible due to the very low level of calcium. Sands have a good diluting effect on phosphorus and potassium, since they are virtually devoid of those two nutrients.

Adding 10–20 percent of hardwood bark fines, pine bark fines, or ground leaf fines gives a nice texture to the growing media. Figure 5A and 5B show the results of adding some bark and leaf fines that were raw and not completely composted. These products result in a generally consistent end product. However, bark fines, leaf fines, or any wood-based material that is not completely composted will have a wide carbon:nitrogen (C:N) ratio, resulting in a nitrogen tie-up and nitrogen deficiency in the plants. Of the wood-based products, pine bark mulch has the highest C:N ratio of wood-based products, followed by wood mulch, hardwood bark mulch, and finally leaf mulch. Table 1 lists the C:N ratios of various composting materials:

Sample Location			Foundry	Cement	Washout	Mason
Sample ID			Sand	Sand	Sand	Sand
Lab Number			69	70	71	72
Sample Depth in inches			9	9	9	9
Total Exchange Capacity (M. E.)			1.15	22.70	16.28	16.86
pH of Soil Sample			9.1	9.1	8.4	8.1
Organic Matter, Percent			0.04	0.51	0.94	0.64
ANIONS	SULFUR:	p.p.m.	7	10	41	70
	Mehlich III Phosphorous:	as (P_2O_5) lbs / acre	41	40	35	40
EXCHANGEABLE CATIONS	CALCIUM: lbs / acre	Desired Value / Value Found / Deficit	454	12512	8424	9328
	MAGNESIUM: lbs / acre	Desired Value / Value Found / Deficit	55	430	582	230
	POTASSIUM: lbs / acre	Desired Value / Value Found / Deficit	25	32	68	45
	SODIUM:	lbs / acre	133	67	54	54
BASE SATURATION %	Calcium (60 to 70%)		65.70	91.88	86.23	92.23
	Magnesium (10 to 20%)		13.27	5.26	9.93	3.79
	Potassium (2 to 5%)		1.86	0.12	0.36	0.23
	Sodium (.5 to 3%)		16.78	0.43	0.48	0.47
	Other Bases (Variable)		2.30	2.30	3.00	3.30
	Exchangable Hydrogen (10 to 15%)		0.00	0.00	0.00	0.00
TRACE ELEMENTS	Boron (p.p.m.)		0.44	< 0.2	0.33	<0.2
	Iron (p.p.m.)		69	57	265	176
	Manganese (p.p.m.)		4	36	194	34
	Copper (p.p.m.)		0.38	0.45	1.6	1.27
	Zinc (p.p.m.)		0.77	0.93	2.55	3.05
	Aluminum (p.p.m.)		47	4	127	3
OTHER	Cobalt ppm		0.026	0.167	1.35	0.279
	Molybdenum ppm		0.04	0.04	< 0.02	<0.02
	Ammonium (p.p.m.)		0.2	0.2	0.3	0.2
	Nitrate (p.p.m.)		1	1.9	1.4	3.4
	Selenium ppm		0.13	0.05	0.15	0.19
	Silicon ppm		18	16.6	30.1	9.8
	Estimated Nitrogen Release #N/Acre		1	21	38	26
	EC mmhos/cm		0.04	0.08	0.17	0.35

Figure 4A: *Standard soil test results for four sands*

			Foundry	Cement	Washout	Mason
Sample Location			Sand	Sand	Sand	Sand
Sample ID						
Lab Number			215235	215236	215237	215238
Water Used			DI	DI	DI	DI
pH			9.1	9.1	8.4	8.1
Soluble Salts		ppm	43	82	151	241
Chloride (Cl)		ppm	10	9	12	19
Bicarbonate (HCO3)		ppm	43	44	44	37
ANIONS	SULFUR	ppm	0.8	4.76	13.38	22.79
	PHOSPHORUS	ppm	0.07	0.07	0.05	0.04
SOLUBLE CATIONS	CALCIUM	ppm	3.53	15.84	32.40	67.18
		meq/l	0.18	0.79	1.62	3.36
	MAGNESIUM	ppm	0.53	3.69	6.89	3.87
		meq/l	0.04	0.31	0.57	0.32
	POTASSIUM:	ppm	0.57	1.46	1.03	2.07
		meq/l	0.01	0.04	0.03	0.05
	SODIUM	ppm	10.20	3.32	3.30	0.97
		meq/l	0.44	0.14	0.14	0.04
PERCENT	Calcium		26.00	61.78	68.53	88.91
	Magnesium		6.45	23.99	24.28	8.54
	Potassium		2.18	2.97	1.13	1.42
	Sodium		65.36	11.26	6.06	1.12
TRACE ELEMENTS	Boron (p.p.m.)		0.03	0.06	0.05	0.02
	Iron (p.p.m.)		3.6	9.1	4.69	0.69
	Manganese (p.p.m.)		0.04	0.28	0.1	<0.02
	Copper (p.p.m.)		< 0.02	< 0.02	< 0.02	<0.02
	Zinc (p.p.m.)		< 0.02	< 0.02	< 0.02	<0.02
	Aluminum (p.p.m.)		1.68	3.24	1.82	0.43
OTHER						

Figure 4B: *Paste test results for four sands*

Table 1: C:N ratios of common composting materials, from Homestead on the Range[1]

Material	C:N ratio
Swine manure	6:1
Aged chicken manure	7:1
Hairy vetch	11:1
Fresh-cut alfalfa	12:1
Table/kitchen scraps	15:1
Used poultry bedding	15:1
Fresh cattle manure	15:1
Sheep manure	15:1
Legume hay	17:1
Fresh grass clippings	20:1
Coffee grounds	20:1
Clover	23:1
Horse manure	25:1
Vegetable scraps	25:1
Mature alfalfa hay	25:1
Wood ashes	25:1
Rye cover crop, vegetative	26:1
Freshly pulled weeds	30:1
Garden waste	30:1
Used horse bedding	45:1
Peat moss	60:1
Leaves	60:1
Fresh cornstalks	60:1
Oat straw	70:1
Wheat straw	80:1
Pine needles	80:1
Rye straw	82:1
Shredded newspaper	175:1
Hardwood bark	223:1
Sawdust	325:1
Shredded cardboard	350:1
Wood chips	400:1
Softwood bark	496:1

---

[1]. Michelle Lindsey, "C:N Ratios of Common Organic Materials," https://homesteadontherange.com/2018/08/27/cn-ratios-of-common-organic-materials. Reprinted with permission. Copyright Homestead on the Range. All rights reserved.

Sample Location			Bark	Leaf	Bark/Leaf	Pine Bark
Sample ID			Fines	Fines	50/50 Mix	Fines
Lab Number			1	2	3	4
Sample Depth in inches			6	6	6	9
Total Exchange Capacity (M. E.)			10.62	5.63	6.85	4.58
pH of Soil Sample			5.3	5.7	5.7	5.9
Organic Matter, Percent			>20	>20	>20	>20
ANIONS	SULFUR:	p.p.m.	16	18	14	7
	Mehlich III Phosphorous:	as (P_2O_5) lbs / acre	158	197	208	108
EXCHANGEABLE CATIONS	CALCIUM: lbs / acre	Desired Value				
		Value Found	1690	940	1197	1230
		Deficit				
	MAGNESIUM: lbs / acre	Desired Value				
		Value Found	257	216	241	344
		Deficit				
	POTASSIUM: lbs / acre	Desired Value				
		Value Found	544	489	576	431
		Deficit				
	SODIUM:	lbs / acre	38	28	29	87
BASE SATURATION %	Calcium (60 to 70%)		39.78	41.76	43.66	44.74
	Magnesium (10 to 20%)		10.08	15.99	14.65	20.86
	Potassium (2 to 5%)		6.57	11.14	10.77	8.04
	Sodium (.5 to 3%)		0.77	1.07	0.92	2.75
	Other Bases (Variable)		6.80	6.00	6.00	5.60
	Exchangable Hydrogen (10 to 15%)		36.00	24.00	24.00	18.00
TRACE ELEMENTS	Boron (p.p.m.)		0.68	0.77	0.77	0.68
	Iron (p.p.m.)		74	15	35	27
	Manganese (p.p.m.)		43	30	35	13
	Copper (p.p.m.)		0.24	< 0.2	0.24	<0.2
	Zinc (p.p.m.)		2.31	1.77	2.4	2.63
	Aluminum (p.p.m.)		31	9	16	30
OTHER	Cobalt ppm		0.062	< 0.02	0.033	0.037
	Molybdenum ppm		0.14	0.09	0.05	<0.02
	Ammonium (p.p.m.)		0.2	0.3	1.6	0.2
	Nitrate (p.p.m.)		5.5	3.9	4.1	2.1
	Selenium ppm		0.16	0.25	0.34	0.2
	Silicon ppm		11.8	9.9	16.4	6.3
	Estimated Nitrogen Release #N/Acre		N/A	N/A	N/A	N/A
	EC mmhos/cm		0.44	0.25	0.33	0.16
	Media Density g/cm3		0.12	< 0.1	0.13	0.1

Figure 5A: *Standard soil test results for some bark and leaf fines*

			Bark	Leaf	Bark/Leaf	Bark/Leaf
Sample Location			Fines	Fines	50/50 Mix	Bark/Leaf
Sample ID						
Lab Number			209386	209387	209388	209389
Water Used			DI	DI	DI	DI
pH			5.3	5.7	5.7	5.9
Soluble Salts		ppm	362	227	238	117
Chloride (Cl)		ppm	30	97	67	95
Bicarbonate (HCO3)		ppm	78	83	66	60
ANIONS	SULFUR	ppm	1.06	1.65	1.1	2.94
	PHOSPHORUS	ppm	2.33	7.99	4.52	2.96
SOLUBLE CATIONS	CALCIUM	ppm	40.15	18.12	22.06	8.18
		meq/l	2.01	0.91	1.10	0.41
	MAGNESIUM	ppm	25.77	15.79	17.02	7.11
		meq/l	2.15	1.32	1.42	0.59
	POTASSIUM:	ppm	55.83	48.76	45.02	27.31
		meq/l	1.45	1.27	1.17	0.71
	SODIUM	ppm	1.12	1.05	0.74	2.77
		meq/l	0.05	0.05	0.03	0.12
PERCENT	Calcium		35.51	25.64	29.63	22.35
	Magnesium		37.98	37.24	38.10	32.34
	Potassium		25.65	35.84	31.41	38.74
	Sodium		0.86	1.29	0.87	6.57
TRACE ELEMENTS	Boron (p.p.m.)		0.41	0.46	0.39	0.1
	Iron (p.p.m.)		1.18	0.9	1.09	0.23
	Manganese (p.p.m.)		5.01	3.07	3.17	0.37
	Copper (p.p.m.)		< 0.02	< 0.02	< 0.02	<0.02
	Zinc (p.p.m.)		0.06	0.04	0.04	<0.02
	Aluminum (p.p.m.)		1.03	0.51	0.71	0.38
OTHER						

Figure 5B: *Paste test results for some bark and leaf fines*

The goal is to end up with a growing medium with an overall carbon:nitrogen ratio less than 24:1. Anything much higher than 24:1 and the microbes in the media mix will pull nitrogen away from your plants and create a nitrogen deficiency. Nitrogen is what drives the boat when it comes to growth and production. When nitrogen becomes deficient, many other nutrients will also become deficient. This will be discussed later in the nutrient section. A blend of bark and leaf fines will create a more balanced and consistent compost mix. As these materials continue to degrade, carbon dioxide will be given off and will stimulate plant growth by serving as a food source for photosynthesis.

The aforementioned composts should make up 75–80 percent of your bulking mix. Secondary bulking agents can make up 15–20 percent of your total mix. These can include products like worm castings, rice hulls, biochar, zeolite, vermiculite, perlite, and pumice. Here is a brief summary of some products that you might use:

Rice hulls

Rice hulls are what's left after processing rice for human consumption. They are a fairly stable, organic product that will provide drainage and aeration for your mix. Another important characteristic of rice hulls is their naturally occurring high level of silicon. Organic mixes are drastically low in silicon, and this is an easy way to not only get silicon, but also improve aeration. Silicon will be discussed further in the mineral section.

Biochar

Biochar is a carbonized biomass which is heated to a charcoal-type consistency, resulting in resistance to degradation. It is lightweight, sterile, has the ability to retain water as well as nutrients, and can provide a habitat for microbial organisms. An inclusion rate of 3–5 percent by volume is considered optimal. Utah State University Plant Health Extension has a nice informational piece on Biochar.[2]

---

[2]. Marion Murray, "What Is Biochar and How Is It Used?" January 2021, https://extension.usu.edu/planthealth/research/biochar.

Zeolite

Zeolite is a natural aluminosilicate which has absorptive capabilities for cations, heavy metals, and moisture up to 60 percent of its weight. In a growing medium which has virtually no exchange sites, zeolite—like biochar—provides exchange sites that stabilize the nutrients in your mix and slowly release them back into the soil solution. A rate of 4–6 percent by volume will have a positive improvement on your bulking mix.

Perlite

Perlite is a naturally occurring volcanic glass-type substance that, when heated to nearly 1600 degrees Fahrenheit, pops like popcorn and becomes a lightweight product that is pH neutral and will absorb water. It is silicon dioxide, similar to sand or quartz. It is oftentimes confused with Styrofoam which, unlike perlite, is purely a bulking agent put in the mix to lower the expense of the end product. Styrofoam doesn't degrade very easily—unlike composts, which will slowly break down over time. After a few years, as the organic material continues to break down into finer particle sizes, this will result in slower drainage, and adding more perlite or vermiculite at 5–15 percent would be justified to improve drainage and aeration. Adding up to 30 percent perlite has been recommended for use in heavier clay soils. When adding perlite to a mix, be sure to wear a mask since the dust will be like breathing in tiny pieces of glass, damaging your lungs.

Pumice

Pumice is another volcanic product primarily used to improve soil structure, drainage, and water holding capacity. It is probably more durable than perlite or vermiculite, but could be used in place of perlite or vermiculite. Because pumice is heavier than the primary bulking agents, it may be difficult to get a nice, even mix.

Vermiculite

Vermiculite is a sterile mica-type product added to mixes to improve drainage, water absorption, and nutrient adsorption. Vermiculite does much the same thing as perlite and pumice. It is pH neutral and might mix better than pumice since it is very light, similar to most of the bulking material. While vermiculite will probably absorb more water than perlite, its usage rate and dust hazards are similar to perlite.

All these products have their place in creating a good growing medium, but I would start off with 5 percent each by volume of rice hulls, biochar, and zeolite. This, along with 10–20 percent manure mulch, 25–35 percent peat, and 40–45 percent compost made from a consistent source of leaf mulch and soft and hardwood bark mulch, would be a great start for making your growing media. I'm quite sure that there are other good substitutes for the bulking products I have mentioned. It all comes down to what's available in your area at the right price and how well it can be balanced in the final mix.

Chapter 2:

Analysis Needed for a Successful Growing Medium

Insert soil with plants.
(Grow Ohio)

This section will discuss the ideal analysis to order from a soils laboratory to provide enough information to create the best possible growing medium. Figures 6A and 6B are examples of all the parameters I think are needed to create a superior product. If you are not building your own growing mix, these tests should be run on a sample of what you want to buy. However, if you test the product you are considering

Sample Location	Compost
Sample ID	10/18/24
Lab Number	87
Sample Depth in inches	6
Total Exchange Capacity (M. E.)	6.06
pH of Soil Sample	6.2
Organic Matter, Percent	>20

			Compost
ANIONS	SULFUR:	p.p.m.	12
	Mehlich III Phosphorous:	as (P_2O_5) lbs / acre	132
EXCHANGEABLE CATIONS	CALCIUM: lbs / acre	Desired Value / Value Found / Deficit	1280
	MAGNESIUM: lbs / acre	Desired Value / Value Found / Deficit	259
	POTASSIUM: lbs / acre	Desired Value / Value Found / Deficit	467
	SODIUM:	lbs / acre	64
BASE SATURATION %	Calcium (60 to 70%)		52.80
	Magnesium (10 to 20%)		17.81
	Potassium (2 to 5%)		9.88
	Sodium (.5 to 3%)		2.29
	Other Bases (Variable)		5.20
	Exchangable Hydrogen (10 to 15%)		12.00
TRACE ELEMENTS	Boron (p.p.m.)		1.37
	Iron (p.p.m.)		42
	Manganese (p.p.m.)		22
	Copper (p.p.m.)		0.23
	Zinc (p.p.m.)		2.25
	Aluminum (p.p.m.)		25
OTHER	Carbon %		25.03
	C/N		30.16
	Cobalt ppm		0.036
	Molybdenum ppm		0.08
	Ammonium (p.p.m.)		0.9
	Nitrogen ppm		8300
	Nitrate (p.p.m.)		1.6
	Selenium ppm		0.12
	Silicon ppm		6.1
	Meida Density g/cm3		0.25

Figure 6A: *Standard soil test results for an example compost*

			Compost
Sample Location			
Sample ID			10/18/24
Lab Number			223815
Water Used			DI
pH			6.2
Soluble Salts		ppm	414
Chloride (Cl)		ppm	100
Bicarbonate (HCO3)		ppm	168
ANIONS	SULFUR	ppm	6.5
	PHOSPHORUS	ppm	6.92
SOLUBLE CATIONS	CALCIUM	ppm	44.17
		meq/l	2.21
	MAGNESIUM	ppm	26.29
		meq/l	2.19
	POTASSIUM:	ppm	49.45
		meq/l	1.28
	SODIUM	ppm	18.09
		meq/l	0.79
PERCENT	Calcium		34.13
	Magnesium		33.86
	Potassium		19.85
	Sodium		12.16
TRACE ELEMENTS	Boron (p.p.m.)		0.62
	Iron (p.p.m.)		0.59
	Manganese (p.p.m.)		2.19
	Copper (p.p.m.)		< 0.02
	Zinc (p.p.m.)		0.05
	Aluminum (p.p.m.)		0.81
OTHER			

Figure 6B: *Paste test results for an example compost*

buying, it doesn't mean that the growing mix will be the same when you go back to purchase more at a later date if it is a different lot or batch.

Most of the commercial mixes that I see are too hot or salty, as a result of adding excess minerals or manure. If you can purchase the bulking components of growing media from a commercial producer, you will be better off if you ask for a limit of 10–15 percent manure compost and then add your own mineral mix. Whether you buy a bulking mix or make your own, Figures 6A and 6B show the parameters you need to get from a laboratory. I use Logan Labs in Lakeview, Ohio, and this is what they refer to as a "complete test with extras." It includes a standard soil test with four extra metals, along with a paste test and an available nitrogen test. Initially, you will also want to add carbon:nitrogen ratio. These tests normally cost less than one hundred dollars.

Once you have the lab data, have calculated the necessary amendments needed, and have incorporated them thoroughly, running at least a follow-up saturated paste test is desirable. It is best if you wait two to three weeks before retesting to allow the mix to reach some kind of equilibrium. Don't expect all the nitrogen to show up on the available nitrogen test, since it is a potassium chloride extraction and will only pick up ammonium (NH_4^+) and nitrate (NO_3^-) nitrogen. Protein nitrogen sources such as blood meal or feather meal cannot be detected until microbes break them down into nitrate or ammonium.

Based on the standard test analysis, the mix evaluated in Figure 6A looks pretty good from an overall perspective. On the major nutrient side, the calcium, sulfur, and phosphorus appear to be low, while on the trace element side, manganese, copper, and zinc appear to be low. Silicon is low, but that is to be expected in a highly organic environment. The carbon:nitrogen ratio is too high, meaning that nitrogen will be pulled away from the plants by microbes if additional nitrogen is not added to the mix to bring the C:N ratio in line (< 24:1). The nitrate and ammonium levels are very low due to microbes consuming nitrogen in an attempt to balance the C:N ratio.

Adjusting a C:N ratio is not difficult if you have all the data necessary, which we do in Figure 6A. Most growing mixes with bulk densities of 0.25–0.35 g/cm³ will weigh between 600 and 700 pounds per cubic yard. In this example, we are going to assume 650 pounds for a cubic yard of mix. The carbon:nitrogen ratio on the test is 30.16:1, and we would like to bring that down to 20:1. Dividing the carbon content of the mix (25.03 percent, which is equal to 250,300 ppm) by 20, we end up with needing 12,515 ppm of nitrogen to get a 20:1 C:N ratio. This mix has 8300 ppm of nitrogen, so 12,515 ppm minus 8300 ppm equals 4215 ppm of nitrogen needed to correct the ratio. If a cubic yard of mix weighs 650 pounds, then multiplying 650 pounds times 4215 ppm, or 0.004215, tells us that we need to add 2.74 pounds of actual nitrogen to the mix to correct the C:N ratio. Assuming that we want to use blood meal as the nitrogen source, which is 13 percent actual nitrogen, then we need to divide 2.74 pounds by 0.13, which means that we need to add 21.1 pounds of blood meal to each cubic yard of mix. This is only the necessary nitrogen for correcting the C:N ratio. It is not the nitrogen required for the growing crop. That will be added on top of the 21.1 pounds of blood meal.

The paste test (Figure 6B) tells quite a different story. As bulking composts go, this would make a great starter mix, since the soluble salts are around 400 ppm and the soluble major nutrients of phosphorus, calcium, magnesium, and potassium are quite good, as are the trace elements boron and manganese. Zinc is on the low side but tolerable. The copper level is below the detection limit but with the high level of organics, most of the copper will be chelated and unavailable. Much of the good soluble levels can be attributed to a great pH level of 6.2. Overapplication of lime or excess salts will result in high pH levels and lower solubility. Growing in media will most likely require a foliar feeding program of at least copper and silicon. I will discuss each of these parameters in the soil chemistry section.

Lab Analysis Explanation

As mentioned earlier, the key analyses needed for optimizing your growing medium are:

1. A standard soil test, which includes available nitrogen, molybdenum, and silicon
2. A paste analysis
3. Carbon:nitrogen (C:N) ratio

The extracting solution for the standard soil test that I will be referring to is the industry standard, called Mehlich-3. This solution has a pH of 2.5 and can easily dissolve any lime in the mix, which will give elevated levels of calcium in the results on the report. For this reason, you cannot depend only on a standard soil test for evaluating your growing media. The pH of a sample will be the result of a 1:1 soil-to-water mix.

The available nitrogen analysis is a potassium chloride (KCl) extraction, which will measure only nitrate (NO_3^-) and ammonium (NH_4^+). Nitrogen in a protein source such as blood meal or feather meal will not be picked up until the microbes break down the proteins into nitrate or ammonium.

Silicon is a separate extraction using acetic acid.

The paste analysis is a distilled water extraction, which is done after holding the sample for 24 hours in a saturated condition. This analysis better represents what nutrients the plants will have access to and will correlate much closer to a tissue analysis than a standard soil analysis.

Carbon:nitrogen ratio is measured by ignition in a carbon/nitrogen analyzer.

Table 2: Desired test values for growing media

	Parameter	Standard test value*	Paste test value
1	Total exchange capacity	6–12 meq/100 g	N/A
2	pH	5.8–6.5	5.8–6.5
3	Organic matter	>20%	N/A
4	Sulfur	25–50 lb/ac	5–15 ppm
5	Phosphorus (P_2O_5)	150–400 lb/ac	2.5–4.0 ppm
6	Calcium	1,200–1,800 lb/ac	50–120 ppm
7	Magnesium	300–450 lb/ac	30–50 ppm
8	Potassium	450–800 lb/ac	40–70 ppm
9	Sodium	50–100 lb/ac	25–50 ppm

10	Base saturation—Ca	50–60%	40–60%
	Base saturation—Mg	10–20%	15–25%
	Base saturation—K	5–10%	15–20%
	Base saturation—Na	<5%	<10%
11	Exchangeable H	5–15%	N/A
12	Boron	1.0–1.5 ppm	0.1–0.3 ppm
13	Iron	>75 ppm	>0.50 ppm
14	Manganese	40–60 ppm	0.10–0.20 ppm
15	Copper	2–4 ppm	0.05–0.08 ppm
16	Zinc	20–40 ppm	0.1–0.2 ppm
17	Aluminum	0–800 ppm	
18	C:N ratio	<20.0	
19	Cobalt	0.04–0.06 ppm	
20	Molybdenum	0.04–0.06 ppm	
21	Selenium	N/A	
22	Silicon	40–60 ppm	
23	Electrical conductivity	0.6–1.0 dS/m	
24	Media density	0.2–0.4 g/cm^3	
25	Soluble salts	N/A	400–600 ppm
26	Chloride	N/A	100–200 ppm
27	Bicarbonate	N/A	90–150 ppm
28	Nitrates (NO$_3^-$)†	15–75 ppm	15–75 ppm
29	Ammonium (NH$_4^+$)	0.1–5.0 ppm	0.1–5.0 ppm

*Crop-dependent
†Based on a 6-inch sampling depth

Parameter description for Table 2

1. **Total exchange capacity (TEC)** for growing media does not mean the same thing as it does for a normal soil. Growing media have virtually no exchange sites for cations. A minor exception would be media which have biochar or zeolite incorporated into the mix. Basically, a growing medium is a pool of organics with minerals suspended in the mix. TEC is calculated by adding the

cations from the Mehlich-3 extraction solution. This solution has a pH of 2.5 and is capable of dissolving minerals, especially lime, resulting in elevated TEC values. Some of these cations may not be available to the plants for some time. For example, the sample from the compost using the old manure mulch shows a TEC of 16.83 meq/100 g and high levels of calcium (2999 lb/ac) in the standard test (Figure 1A), but relatively low levels of calcium in the paste solution (Figure 1B). This is not to say that the calcium won't eventually become available, but it emphasizes the need for a paste test to show exactly what level of nutrients will be immediately available to the plants. The ideal TEC should be between 6.0 and 12.0 meq/100 g. TEC values over 12.0 meq/100 g indicate possible excess nutrients and an increased risk of excessive salts. TEC values over 20 are almost certain to have both high pH and excessive salts.

2. **pH** of growing media should be between 6.0 and 6.5. The pH could go as low as 5.8, as long as the soluble cation numbers are met. A pH higher than 6.5 with excess calcium and magnesium will limit the availability of phosphorus and trace elements. High levels of salts will cause the pH to exceed 6.5 by blocking the hydrogen ions from getting detected by the pH electrode. This may not hurt the solubility of phosphorus and trace elements if the salt happens to be sodium. pH levels below 5.5 will have a negative impact on the biological system, causing a shift from bacterial dominance to fungal dominance.

3. **Organic matter,** by nature of the composition and definition of growing media, will almost always test at the lab's optimum organic matter reading over 20%.

4. **Sulfur** levels should be 25–50 ppm on the standard soil test and 5–15 ppm on the paste test. Sometimes sulfur levels will go much higher than the optimum, especially if too much gypsum or potassium sulfate is added. Most of the potassium should be coming from the composted plant material and/or manure, minimizing the need to use much potassium sulfate.

5. **Phosphorus** is reported on the standard soil test as P_2O_5. To con-

vert those numbers to elemental phosphorus (P), divide by 4.6 when sampled at 6 inches and 6.9 when sampled at 9 inches. The phosphorus level of 132 lb/ac in Figure 6A of the standard test is relatively low by my standards in Table 2; however, the paste test level of 6.9 ppm is quite good. The results of the paste test will be paramount to the plant on a short-term basis, but the results of the standard test should not be ignored. The levels in the standard test will provide the feed rate for future uptake by the plants. In this case, there is plenty of soluble phosphorus, but adding some low-solubility rock phosphate to the mix would be a good idea. I will discuss these situations further in the section on amending growing media (Chapter 5). For established beds or pots, a phosphorus range of 150–400 lb/ac or 32–90 ppm on the standard test and 2.5–3.5 ppm on a paste test should supply an adequate amount of short-term phosphorus as well as long-term phosphorus. I have seen values over seven times my normal range of 150–400 lb/ac, but they do not meet the paste requirements of 2.5–3.5 ppm due to excess calcium or magnesium and a high pH. Even if the calcium is coming from gypsum, which is very soluble, the free calcium can precipitate phosphorus out of the solution into a low-solubility form of rock phosphate. Well water high in calcium is another source that will precipitate out phosphorus into rock phosphate.

6. **Calcium** levels on the standard test should be between 1,200 and 1,800 lb/ac or 600–900 ppm. I consider most of this calcium to be reserve or replacement for the plant-available calcium in solution. The paste numbers are more the priority and should range from 50–120 ppm. I would use the higher side of the range for things like greens, tomatoes, and cannabis. Greens should do well at 50–60 ppm, while tomatoes might prefer a little higher level of 70–80 ppm. Cannabis would like the highest level of calcium, at 100–120 ppm. Any of the nutrient levels could vary up or down depending on the feed rate from the reserve nutrients to the soluble portion. Plant requirements for nutrients will also go up or down depending on the stage of growth.

7. **Magnesium** on the standard report should have a ratio of 4–5 parts calcium to 1 part magnesium. Therefore, magnesium on the standard report should be approximately 300–450 lb/ac or 150–225 ppm if the calcium is 1,200–1,800 lb/ac. On the paste test, I would like to see the Ca:Mg ratio a little closer, at 2:1–3:1. Be careful when balancing by using ratios, especially if one of the cations has been applied in excess. For example, if calcium tested out at 3,600 lb/ac on the standard soil test and you are using a Ca:Mg ratio of 4:1, then ideally the magnesium should be set at 900 lb/ac. Making that kind of adjustment could easily get you into a soluble salts issue. When running into that kind of issue, I would work on balancing the paste test and not the standard soil report.

8. **Potassium** I would also balance using a potassium:magnesium ratio of 1.5:1–2.0:1. On the standard soil report, if magnesium is running 300–400 lb/ac, then ideally the potassium should range around 450–800 lb/ac, or 225–400 ppm. Potassium and magnesium are a direct line of interference with each other, so getting too far out of line with each other will induce deficiencies between them. Balancing magnesium against potassium using ratios is fine until one of the cations is in excess, and then soluble salts may become an issue.

9. **Sodium** between 50–100 lb/ac or 25–50 ppm is a good target for the standard soil test. A good rule of thumb is to never get your sodium base saturation percentage higher than your potassium base saturation percentage. This also applies to the paste test. On the paste test, try to keep the sodium under 50 ppm and the base saturation less than 10. High levels of sodium in growing media will not lead to soil structural problems like in a normal soil but will lead to high pH and interference issues. A high pH caused by sodium will not decrease phosphorus availability like a high pH caused by excess calcium. Sodium phosphates are very soluble, while calcium phosphates are not. Manure additives and sea-based products can be big sources for sodium, along with some irrigation waters.

All of these desired numbers are dependent upon the coarseness and solubility of the products being used in the mix. The goal is to hit the paste numbers more than the standard soil test numbers. However, keep in mind that the standard soil test numbers are the feed supply for the paste numbers. If your cation amendments are coarser and possibly lower in solubility, then the standard soil test numbers will have to be elevated to get the desired paste numbers. When it comes to lime, a 100–200 mesh size would be preferable. Don't over-lime and get the pH above 6.5. At pH levels 7.0 and above, the solubility of the lime, regardless of the particle size, will drop to about 10 percent.

It is best to find as many of the minerals as you can with the same particle size, somewhere between the size of flour and sugar consistency.

10. **Base saturation** numbers in growing media are less important than in a clay-based soil, where you are trying to use the Albrecht philosophy of balancing soils to improve soil physics. With that said, I would still prefer having more calcium than magnesium, more magnesium than potassium, and more potassium than sodium on a percentage basis. A calcium base saturation range of 50–60 percent, magnesium range of 10–20 percent, potassium range between 5–10 percent, and a sodium saturation less than 5 percent would be ideal for the standard soil test. The paste test mirrors those numbers pretty closely, with calcium 40–60 percent, magnesium 15–25 percent, potassium 15–20 percent, and sodium less than 10 percent. The finer and more soluble your amendments are, the lower the range in the standard test should be.

11. **Exchangeable hydrogen** should range between 10–15 percent. This is a bit of misnomer, since growing media really don't have much in the way of exchangeable sites; however, a certain amount of hydrogen is needed to get the pH to 6.2–6.5. If the pH is 7.0 or above, the exchangeable hydrogen ions will be zero. This does not mean that you have no hydrogen ions in the mix; if you did the pH would be 14. At pH levels above 7, it is expected that any hydrogen ions on the colloids would be replaced by the more

dominant cations such as calcium and magnesium.

12. **Boron** on the standard test should range between 1.0–1.5 ppm. The paste test should range between 0.1–0.3 ppm. A lower range on the standard test could provide an adequate range in the paste test, depending on the amount of excess calcium and magnesium in the mix. Calcium and magnesium borates are not water-soluble. Five parts per million of boron on the standard test and 2.0 ppm in the paste test will subject your plants to boron toxicity.

13. **Iron** should be greater than 75 ppm on the standard soil test and greater than 0.25 ppm on the paste test. Iron will be more soluble at lower pH levels. Some compost windrowed on soil may have small amounts of clay in the mix. Should any of the fine clay particles slip past the filter paper during filtration at the lab, this will show up in the inductively coupled plasma (ICP) data and be translated as high iron and aluminum, the main components of clay. This can happen in both the standard and paste test. This is the same result that I see on tissue tests that might get some soil dust on them while sampling. I am more worried about low iron than some elevated numbers on the standard or paste tests. Iron and aluminum become more soluble at pH levels below 5.8. Most of the time if you fix the low pH, the iron and aluminum numbers will come into line.

14. **Manganese** availability is very pH dependent. I have seen field soil reports with readings of 10 ppm manganese, which by anyone's standard is very low. In one particular muck field, the north half of the field of carrots exhibited manganese deficiency—dark green veins and yellow interveinal tissue—while the south half of the field exhibited manganese toxicity—dark spots on the leaves. How could a field with both ends having 10 ppm manganese be this dramatically different? The answer was soil pH. The north half of the field, where the rich black organic soil was playing out, resulting in calcareous material being incorporated by tillage, had a soil pH of 7.5 and the south end, where the organic soil was deeper, had a soil pH of 4.8. Muck soils, like growing media, are rather light and fluffy with good aeration, which

increases the oxidation of manganese into an unavailable form. Therefore, setting a hard-and-fast number for manganese on the standard soil report is difficult at best. However, if you keep the soil pH between 6.0 and 6.5, a value of 40–60 ppm will give you the best chance for maximum availability. The paste test will be the more reliable of the two tests. Levels between 0.07 and 0.15 ppm should provide enough available manganese.

15. **Copper** levels on a standard soil report of 5–10 ppm might work in a normal soil, but it probably won't work in a highly organic growing medium since the organic constituents tend to chelate copper quite well. A copper level on the standard soil report of 2–4 ppm is good, but having the paste test in the range of 0.05–0.08 ppm is ideal. Since copper is so easily chelated, a tissue test would be the best way to verify, but also consider a foliar feeding program during the growing season.

16. **Zinc** levels on the standard soil test of 20–40 ppm, along with paste values between 0.1 and 0.2 ppm, are ideal. It will be difficult to reach these zinc levels due to phosphorus tie-up. Phosphates precipitate out zinc into a very low-solubility zinc phosphate. Excess manure in the compost only tends to exacerbate this problem. Foliar feeding zinc along with copper will circumvent the interference issues and directly feed the plants.

17. **Aluminum** could essentially be zero on both the standard soil report and the paste analysis. It is a non-essential element found in the earth's crust in high amounts. High levels of soluble aluminum can interfere with uptake of other mineral elements. Solubility of aluminum occurs at low pH levels and is unlikely to occur if normal pH levels are maintained. Like iron, aluminum is part of the clay lattice structure, and clays are extremely fine particles. If any of these particles get past the filter paper during the filtration of the standard or paste test, they will show up as elevated levels of iron and aluminum. It is also possible to have fine clays in well and municipal water supplies. Toxic levels of aluminum have more to do with interference and tie-up of essential nutrients. This might occur with greater than 1500 ppm on

the standard soil report and a pH level well below 6.

18. **Carbon:nitrogen** ratio for growing media should be below 20:1. Higher levels will result in microbes tying up nitrogen in order to establish a lower ratio. This nutrient deficiency is easily fixed by adding more nitrogen, as long as the C:N ratio is not too wide. Note the C:N ratios of various organic sources from Home on the Range in Table 1 (Page 13).
19. **Cobalt** range on the standard soil test of >0.06 ppm should be adequate. Detection limit at the lab is 0.02 ppm.
20. **Molybdenum** levels of 0.04–0.08 ppm should be adequate. Detection limit is 0.02 ppm.
21. **Selenium** range is undecided.
22. **Silicon** range on the standard soil test should be between 40–60 ppm.
23. **Electrical conductivity (EC)** is measured in the standard soil test protocol. Ideal values are <0.6 dS/m for direct seeding and <1.1 dS/m for transplants. These are good rule-of-thumb numbers, but these values could fluctuate up or down depending upon the plant species.
24. **Media density** is a comparison number with no real desired value. It can be used to compare the weight of a growing medium to a normal soil. This number is what I use to determine how much of an amendment should be used to improve the balance of a mix. It is very easy to over-fertilize and create a salt issue without using this number. Growing media with a lot of perlite or Styrofoam beads will have bulk densities from 0.15 to 0.25 g/cm^3. A lot of softwood bark mulch or leaves will also lighten the density. Typically, growing media with various composts as their bulking agents will range between 0.25 to 0.35 g/cm^3. Low bulk densities generally require more watering, resulting in more leaching.
25. **Soluble salts** is a measurement of all cations and anions in the paste extract. It is not just a sodium issue. Soluble salts and electrical conductivity (EC) are both measured with an electrical conductivity meter. The only difference between EC and soluble salts is the length of time between readings. The EC reading on

a standard soil test is read within minutes of adding the distilled water, while the EC reading on the paste test is done after the 24-hour holding time. The soluble salts reading will almost always be higher than the EC reading in the standard soil test because it has a longer time to equilibrate, allowing more soluble salts to come into solution. Given a choice, I always base my decisions on the paste test.

This is the number one problem that comes across my desk when dealing with growing media. Soluble salts will be discussed more thoroughly in the amendment and irrigation water sections (Chapters 3 and 4).

Germinating seeds do best with soluble salt levels less than 400 ppm or 0.6 dS/m. Larger transplants will tolerate up to 800 ppm or 1.25 dS/m, but may start a little slowly as they adjust to the higher level of salts. Based on the problems that I see coming across my desk, 800–1,000 ppm is the maximum salt level in growing media before we start seeing burning on leaf edges, yellow leaves on the bottom of the plant, and premature drought stress, to name a few. This is a fairly conservative upper limit compared to what other growing media formulators have told me.

The factor a lot of people forget to enter into the soluble salts equation is the level of salt in their irrigation water. If you are watering with rain or reverse osmosis (RO) water, salts are not an issue. However, well or municipal water sources can have a wide range of levels and will quickly complicate the soluble salts issue.

26. **Chloride** is an anion nutrient that plants take up quite readily to balance the electrical charge of the various cations taken up. A range of 100–200 ppm would be acceptable to maintain the electrical balance in the plant. Anything much higher will contribute to a soluble salts issue.

27. **Bicarbonate** is an anion formed by the dissolution of calcium carbonate. A range of 90–150 ppm is ideal. When lime breaks down in soils, bicarbonates are one of the end products. Bicarbonates form a weak acid, while calcium forms a strong base, calcium hydroxide, which causes an increase in the pH of the soil. Cal-

cium and bicarbonates are found in limestone-based well water and can accumulate in your soil through irrigation. As the soils dry down, bicarbonate can reconnect with the calcium to precipitate out pure calcium carbonate, which can react with phosphates and precipitate them out as low–solubility rock phosphate. High levels of bicarbonate can precipitate out iron as a very insoluble iron carbonate, inducing an iron deficiency. Adding sulfuric, acetic, or humic acids will neutralize the bicarbonate, gassing it off as carbon dioxide. Consult with an irrigation specialist before attempting this.

Table 3: Standard soil test desired values for growing media

Parameters	Bed 6" Deep	Bed 12" Deep	Bed 18" Deep
Total exchange capacity	6–12 meq/100 g	6–12 meq/100 g	6–12 meq/100 g
pH	5.8–6.5	5.8–6.5	5.8–6.5
Organic matter	>20%	>20%	>20%
Sulfur	25–50 ppm	15–35 ppm	10–25 ppm
Phosphorus (P_2O_5)	150–400 lb/ac	100–270 lb/ac	60–180 lb/ac
Calcium	1,200–1,800 lb/ac	800–1,200 lb/ac	400–800 lb/ac
Magnesium	300–450 lb/ac	200–300 lb/ac	100–150 lb/ac
Potassium	450–800 lb/ac	350–560 lb/ac	250–320 lb/ac
Sodium	50–100 lb/ac	35–70 lb/ac	20–40 lb/ac
Base saturation—Ca	50–60%	50–60%	50–60%
Base saturation—Mg	10–20%	10–20%	10–20%
Base saturation—K	5–10%	5–10%	5–10%
Base saturation—Na	<5%	<5%	<5%
Exchangeable H	5–15%	5–15%	5–15%
Boron	1.0–1.5 ppm	0.7–1.0 ppm	0.4–0.5 ppm
Iron	>75 ppm	>50 ppm	>25 ppm
Manganese	40–60 ppm	30–45 ppm	20–30 ppm
Copper	2–4 ppm	1.5–3.0 ppm	1.0–2.0 ppm
Zinc	20–40 ppm	15–30 ppm	10–20 ppm
Aluminum	0–800 ppm	0–600 ppm	0–400 ppm
C:N Ratio	<24.0	<24.0	<24.0
Cobalt	0.04–0.06 ppm	0.04–0.06 ppm	0.04–0.06 ppm

Molybdenum	0.04–0.06 ppm	0.04–0.06 ppm	0.04–0.06 ppm
Selenium	N/A	N/A	N/A
Silicon	40–60 ppm	30–50 ppm	20–40 ppm
Electrical conductivity	0.6–0.9 dS/m	0.5–0.8 dS/m	0.4–0.7 dS/m
Media density	0.2–0.4 g/cm^3	0.2–0.4 g/cm^3	0.2–0.4 g/cm^3
Soluble salts	N/A	N/A	N/A
Chloride	N/A	N/A	N/A
Bicarbonate	N/A	N/A	N/A
Nitrates (NO$_3^-$)	15–75 ppm	10–40 ppm	6–25 ppm
Ammonium (NH$_4^+$)	0.1–5.0 ppm	0.1–5.0 ppm	0.1–5.0 ppm

Explanation of Table 3

My thoughts for creating Table 3 are based on soluble salts and the longevity of the bed for repeated growing cycles. If we create an 18-inch bed with the nutrient level of a 6-inch bed, then we will have more nutrients than needed to grow a crop. Therefore, as we irrigate the bed during a growing cycle, some nutrients will move down into the lower depths, creating the potential for excess nutrient or soluble salt buildup. You may not notice the problem for a few growing cycles, especially if you only test the top 6 inches and reamend accordingly. Eventually, the soluble salts level in the bottom portion of the bed will become salty enough to restrict root growth and crop production.

Nutrient movement into the lower level of the bed is inevitable, so monitoring the lower depths of the bed will allow you to catch this problem early, allowing you to deeply mix the bed and not have to remove the mix and start over. This will save a lot of money and work.

Turmeric growing in media-based soil on the Big Island of Hawaii.
(Courtesy of Stephen Filipiak)

Chapter 3:

The Impact of Nutrients on Plant Growth

In this section, we will look at various essential nutrients and their impact on plant growth. We will also look at how these nutrients move in the plant and how deficiency symptoms can help us identify which nutrients may be limiting plant growth. Understanding these factors will help us minimize fertilization and maximize plant growth.

Growing in media is far more complicated than growing in a garden or field. Although we have more control over environmental factors such as light and water, the interactions between nutrients in a medium that has little or no soil exchange sites and no buffering capacity for excesses or deficiencies makes it extremely important that we know and understand the nutrient demands of our growing crop.

Tap Root cukes.
(Tap Root Fields, Ben Zoba)

Much of the following information is taken from my previous book, *A Growers Guide for Balancing Soils: A Practical Guide to Interpreting Soil Tests* (Acres U.S.A., 2021). With this information, you will be a more successful grower and understand what the plants are telling you as they grow.

Before we get into this information in detail, a little explanation of the following topics is in order:

Ideal soil levels—these are levels that I generally feel are adequate for crop production. Of course my levels are based on averages, and as the saying goes, "If you put one foot in hot water and the other in cold water, on the average you won't be comfortable." These numbers from me or anyone else are starting numbers and can and should be fine-tuned to your situation.

Mobility in the plant—a yes or no tells you, once the nutrient gets into the plant, whether it can be remobilized and moved to a higher-priority part of the plant.

Xylem- and phloem-mobile—this tells you how mobile the nutrient is in the plant. Xylem is the water transport mechanism in the plant and it is one-directional—up. As water transpires off the leaf surface, it creates a vacuum, allowing water to be pulled up from the roots. With that water are various nutrients that may get dropped off along the way up into the plant. Once a nutrient is dropped off, the only way of moving back down to a needed area is if it is capable of traveling through the phloem. Nutrients that are only xylem-mobile can only be remobilized to an area higher up in the plant. The phloem is like an elevator that goes up or down. Being phloem-mobile means the nutrient can remobilize anywhere in the plant and be taken to a higher-priority location, whether to the plant canopy or the roots.

Site of the initial deficiency symptoms—tells you whether the nutrient is xylem-mobile, phloem-mobile, or both. Deficiencies that start low in the plant are highly mobile.

Role in the plant—a more detailed way of knowing which nutrients will be in high demand in which part of the plant for your growing crop. For example, if you are growing a crop for protein, nitrogen and sulfur will be highly prioritized nutrients. However, larger plants and higher plant populations will also increase the demand for nitrogen, even if you are growing the crop as a carbohydrate source.

Deficiency symptoms—where they show up on the plant gives a clue as to which nutrients may be deficient. For example, nitrogen deficiency will show up on the bottom of the plant and down the midrib of the leaf in grasses, but potassium will show up at the bottom of the plant and on the outer edges of the leaf.

Toxicity symptoms—assuming the toxicity doesn't kill the plant, understanding the symptoms from excesses will help rein in your nutrient applications. Toxicities can occur from lack of applications such as lime, resulting in very low pH levels and increased solubility of elements such as aluminum.

Understanding these issues will make you a more informed grower. Use soil and/or tissue sampling to make these determinations.

Nitrogen

Nitrogen can be your friend or your enemy, depending how you use it. One thing is for certain—*nitrogen will drive the boat*. There is no other nutrient that has as big of an impact on plant growth as nitrogen. A small deficiency will immediately lead other nutrients into deficiency.

- **Ideal media level:** Crop and soil dependent
- **Mobile in the plant:** Yes
- **Xylem- and phloem-mobile:** Yes
- **Site of initial deficiency symptoms:** Older leaves
- **Role in the plant:** Formation of amino acids, vitamins, and cell division. Major component of plant proteins.
- **Deficiency symptoms:** Yellowing at the base of the plants and moving up as the deficiency grows. In grasses, yellowing will start at the base of the plants, with the yellowing starting at the tip and moving down the midrib of the leaves. Plant growth will slow substantially and leaves will eventually turn completely yellow.
- **Toxicity symptoms:** Very dark green, excess vegetative growth, reduced flowering, lodging, increased disease and insect pressure

Sulfur

Sulfur in growing media is often very high due to the addition of potassium sulfate and gypsum. High sulfates are better than high chlorides or bicarbonates, but they still contribute to the soluble salts and need to be added judiciously. Elemental sulfur used to lower the pH will eventually end up as sulfate. It is important to understand that elemental sulfur is primarily broken down by thiobacillus bacteria, which must be in the media to effectively break down elemental sulfur. When using elemental sulfur, it is important to use the feeding grade, which has grains about the size of sugar. Flowers of sulfur are very fine and dusty, so be sure to wear a mask to avoid getting it in your lungs.

- **Ideal media level:** Standard soil test 25–35 ppm, paste test 5–15 ppm
- **Mobile in the plant:** Minimal
- **Xylem- and phloem-mobile:** Primarily xylem
- **Site of initial deficiency symptoms:** New growth
- **Role in the plant:** Formation of amino acids, enzymes, and vitamins. Sulfur will impact seed development, disease tolerance, level of nitrates, and chlorophyll production.
- **Deficiency symptoms:** Plants will be pale-green to yellow if the deficiency is bad enough. The new growth will show the deficiency first, unlike nitrogen, which starts on the lower/older growth. Young deficient leaves may also exhibit interveinal chlorosis. A sulfur shortage also causes plants to be shorter and squatter than normal. The yellow leaves may also be more upright and narrower.
- **Toxicity symptoms:** Sulfur toxicity is rare, but sulfate burn caused by foliar feeding of too much sulfate-based nutrients can happen quite easily. Start by using less than 4–5 pounds of combination or single sulfated nutrients per one hundred gallons.

Phosphorus

Media often have very high phosphorus levels due to high amounts of manure compost in the mix. Limiting the manure-based compost to 10–15 percent of the mix will help reduce excess phosphorus and potassium.

- **Ideal media level:** Standard soil test 150–400 lb/ac P_2O_5, 30–85 ppm P_2O_5; paste test 2.5–4.0 ppm
- **Mobile in the plant:** Yes
- **Xylem- and Phloem-mobile:** Yes
- **Site of initial deficiency symptoms:** Overall stunted plant size. Usually occurs early in the season.
- **Role in the plant:** Regulates the energy in the plant, cell growth, and root and seed formation; promotes winter hardiness; necessary for carbohydrate and protein synthesis; promotes lignification
- **Deficiency symptoms:** Phosphorus deficiencies mainly inhibit plant growth, resulting in a smaller, darker plant with reddish coloration. The new leaves will be smaller than normal. A phosphorus deficiency will also affect a plant's reproductive performance, resulting in fewer and smaller seeds.
- **Toxicity symptoms:** Excess phosphorus rarely leads to a terminal toxic effect on plants. Excessive levels can exacerbate a deficiency in calcium, boron, iron, manganese, copper, and zinc. Zinc and iron are normally the two nutrients most affected. Phosphorus will be excessive in older tissues of plants if the concentration in whole leaf tissues exceeds 1 percent. Conditions like reduced leaf size in the new growth or premature ripening may occur due to zinc tie-up. If iron becomes substantially affected, plants may exhibit tissue chlorosis.

Calcium

Limestone is the major source of calcium for most mixes. There are two sources of limestone. High-calcium lime is primarily calcium, with 30–34 percent calcium and less than 2 percent magnesium. Dolomitic lime is quite different, with 21 percent calcium and 11 percent magnesium. If you need magnesium to balance the bulking mix, use dolomite or a combination of dolomite and high-calcium lime. Keep the pH levels between 6.2–6.5. Gypsum can be used to add calcium to the mix; because it is pH neutral, it won't raise the soil pH. Calcium silicate, also known as Wollastonite, can also be used to add calcium, but it will raise the pH. Since most organic mixes are deficient in silicates, using calcium silicate for a portion of calcium supplementation is a good idea. Generally, calcium silicate is a low-solubility product, so use the finely ground products.

- **Ideal media level:** Standard soil test 1,200–1,800 lb/ac at 6-inch depth, paste test 50–120 ppm
- **Mobile in the plant:** No
- **Xylem- and phloem-mobile:** Xylem
- **Site of initial deficiency symptoms:** New leaves
- **Role in the plant:** Critical for root development, cell wall integrity, nitrogen metabolism, seed germination, pollination, and fruit set
- **Deficiency symptoms:** Calcium is not mobile in the plant; once incorporated in the plant structure, it cannot be remobilized. Therefore, deficiency symptoms are limited to new growth. Leaves will be small and show possible necrosis on the leaf tip and edges. New leaves will curl down and develop a claw-like appearance. Cereal grains may have reduced tillering and soft and limp stems.
- **Toxicity symptoms:** Calcium toxicity is relatively unheard of. Calcium will raise the pH and affect other nutrients through reduced solubility or direct interference with the uptake of other cations.

Magnesium

Dolomitic limestone, Epsom salts (magnesium sulfate), and sulfate of potash magnesia (K-Mag or Sul-Po-Mag) are the most common sources of magnesium. Try to acquire as much of your magnesium from lime or natural sources as possible. Wood ash is one natural source that contains 3–5 percent magnesium along with 25–45 percent calcium. K-Mag would be my next choice because of the nice balance between potassium and magnesium. Excessive potassium will interfere with magnesium uptake. Plant-based composts and various blends may have significant amounts of potassium, so if extra magnesium is needed, Epsom salts would be the product of choice.

- **Ideal media level:** Standard test 300–450 lb/ac at 6-inch depth, paste test 30–50 ppm
- **Mobile in the plant:** Yes
- **Xylem- and phloem-mobile:** Yes
- **Site of initial deficiency symptoms:** Older leaves, but not necessarily the very bottom leaves.
- **Role in the plant:** Chlorophyll formation, phosphorus movement in the plant, fruit maturity, over 300 enzyme functions, pollen germination
- **Deficiency symptoms:** The visual symptoms almost always begin with chlorosis of the older leaves and then move to the newer growth. For grasses, the deficiency symptoms appear as light striping. Dicots generally have interveinal chlorosis and blotchiness on the older leaves. Reddish to brown margins will occur on some plants, like cauliflower and cabbage. Pines will generally show yellow tips on the needles of the tree, and the yellowing will proceed to the base of the needle. Normally this will start occurring on the lower portions of the plants.
- **Toxicity symptoms:** Excess magnesium, like calcium, doesn't cause toxicity per se but can interfere with the uptake of other cations, especially potassium and manganese

Potassium

A mix with plant-based compost and some limited manure should supply the bulk of the potassium. If a little magnesium is also needed to finish out the balancing, K-Mag would be my first choice. If potassium only is needed, sunflower hull ash (0-4-34) is a great natural source. Potassium sulfate is also a great product which is very soluble, but carries with it a high level of sulfates.

> - **Ideal media level:** Standard test 450–800 lb /ac at 6-inch depth, paste test 40–70 ppm
> - **Mobile in the plant:** Yes
> - **Xylem- and Phloem-mobile:** Yes
> - **Site of initial deficiency symptoms:** Older leaves
> - **Role in the plant:** Carbohydrate metabolism and mobilization, regulation of water utilization through leaf stomata, summer coolant and winter antifreeze, affecting fruit color and sweetness, increasing disease resistance through the metabolism of sugars to carbohydrates
> - **Deficiency symptoms:** Potassium does so much in the plant that a deficiency may have many faces. Deficient plants will grow slowly and eventually become stunted, with shorter internodes. Since potassium is so mobile in the plant, a deficiency will tend to show up on the older leaves. In grasses, a potassium deficiency will show up as burning on the outside edges of the leaves. Other plants will exhibit burning at the leaf tip on the older leaves, with the base of the leaf remaining green. Plants with a potassium deficiency will oftentimes wilt more rapidly in the heat of the day. Plants will tend to lose their lower leaves as the deficiency progresses, leaving crops like alfalfa stemmy and low in protein. Diseases and insect pressure may increase since sugars may accumulate in the tissue of the leaves or roots. Potassium can act as an antifreeze in the plant, helping to resist freezing during the large temperature swings that often occur in the winter. Grain from potassium-deficient plants will be smaller and lighter in test weight.

> Vegetables like tomatoes may be small and off-color. The shoulders of the tomatoes will often be yellow, especially on the later-setting fruit. Greens may be bitter. Bruising is more prevalent on low-potassium fruit, especially potatoes, tomatoes, peppers, and tree fruit. Greens may be bitter. Bruising is more prevalent on low-potassium fruit, especially potatoes, tomatoes, peppers, and tree fruit.
> - **Toxicity symptoms:** Excess potassium will primarily result in driving soil pH levels high and causing a salt burn on the tissue. Excessive potassium applications can interfere with uptake of calcium, magnesium, and ammonium ions, as well as the trace elements of boron, manganese, and zinc.

Trace Elements

It would be nice to be able to supply all the trace elements with rock dusts and products like Azomite, but their solubilities are very low and it is hard to know if the level of trace elements needed at critical times will be there when needed. Supplying at least 50 percent of the trace elements with rock dusts would give you a slow release of these nutrients and reduce the risk of overapplying soluble trace elements and creating a problem that will really be hard to fix.

Boron

Sodium borate can easily be found in the grocery as Borax. The old 20 Mule Team Borax has been around for decades. Most suppliers carry soluble boron, along with granular boron containing both sodium and calcium borate for a measured soil release. Calcium borate is not very soluble.

- **Ideal media level:** Standard soil test 1–1.5 ppm, paste test 0.1–0.3 ppm
- **Mobile in the plant:** No
- **Xylem- and phloem-mobile:** Xylem
- **Site of initial deficiency symptoms:** New growth
- **Role in the plant:** Cell wall formation, pollen grain germination, seed production, sugar translocation
- **Toxicity:** The range between adequate levels and toxic levels can be quite small. The older leaves will show yellowing, necrosis, and blotches starting on the margins of the leaves and spreading into the center of the leaves. Yields and plant size will be depressed. Toxicity on tomatoes will occur at the leaflet tips and curl inward. Flushing out toxic levels of boron will take 3–4 times more water than flushing out sodium chloride.

Iron

Iron is generally not a problem in growing media—or normal soils for that matter. All living things have iron, and as we compost different materials such as manure, bark mulch, and peat, the iron, along with many other elements, tends to concentrate during the composting process. Iron deficiency in growing media is primarily due to excess liming and/or watering with high-bicarbonate water. Once iron is precipitated out as iron carbonate, it is very insoluble.

- **Ideal media level:** Standard soil test >75 ppm, paste test >0.5 ppm
- **Mobile in the plant:** No
- **Xylem- and phloem-mobile:** Xylem
- **Site of initial deficiency symptoms:** New growth
- **Role in the plant:** Chlorophyll formation, oxygen carrier, cell division and growth

> - **Toxicity:** Iron toxicity rarely ever occurs, except sometimes in rice production, where the saturated soil conditions can reduce iron, making it more available. Excess iron could result in phosphorus tie-up in low pH situations, but it is more likely to occur from over-fertilizing with iron.

Manganese

Manganese availability is very sensitive to pH and oxidative conditions. The bottom line is that manganese becomes less soluble as the pH approaches 7 and above. Manganese is most available in a reduced state and the light, fluffy, well-aerated condition of growing media exacerbates the oxidation of manganese, making it less available. Therefore, over-liming and raising the pH is strike one, a well-aerated mix is strike two, and watering with high-bicarbonate water is strike three—you're out. Two of the three strikes are fixable by keeping the overall pH of the mix between 6.2 and 6.5 and, if your water is high in bicarbonates, acidifying your water to a pH of 6.2–6.5. Tissue sampling and having a foliar feeding program in hand will allow you to avoid deficiencies.

> - **Ideal media level:** Standard test 40–60 ppm, paste test 0.10–0.20 ppm
> - **Mobile in the plant:** Semi-mobile
> - **Xylem- and phloem-mobile:** Yes
> - **Site of initial deficiency symptoms:** The leaves midway up the plant, at the new growth. Deficiency symptoms will start higher up on the plant as compared to magnesium deficiency, which will start lower on the plant.
> - **Role in the plant:** Chlorophyll synthesis, many enzyme systems, improving calcium and phosphorus uptake, aiding in disease resistance.
> - **Toxicity:** Manganese toxicity is generally the result of a low-pH soil environment, below 5.5. Levels in leaves above 300–550 ppm are

> high enough to result in toxicity. Some plants may be higher or lower, but numbers in this range are a good indicator that problems may be just around the corner. Symptoms are brown specks on the leaves, browning of stalks, chlorotic lesions on the leaf tip, and crinkling of the leaves. Iron chlorosis could be a secondary issue due to manganese interference. Some plants may be more tolerant to excess levels of manganese in the soil due to their poor ability to absorb manganese. Plants that prefer to grow in higher-calcium soils, such as cannabis, tomatoes, and legumes, will be more sensitive to toxicities.

Copper

Copper is a trace element that is picked up by direct root intercept. Seventy percent of copper and 50 percent of iron is picked up by the plant this way. This means root mass is critical for plants to obtain this nutrient. This is not a problem in growing media since roots can proliferate easily in the mix, unless there is a real calcium shortage or too high of soluble salts. However, barring these two situations, copper uptake into the plants is still a problem. Paste analysis for copper indicates that over 90 percent of the time it is below the 0.02 ppm lab detection limit. The problem is that the high level of organics in growing media tends to chelate copper, walling it off from the plant roots. Using a chelated amendment of copper allows for a timed release of copper so that the plants have a chance to pick the metal up before it is chelated again. Chelated products are not generally accepted as organic; therefore, foliar feeding copper sulfate is about the only option for plant absorption.

> - **Ideal media level:** Standard soil test 2–4 ppm, paste test 0.05–0.08 ppm
> - **Mobile in the plant:** No
> - **Xylem- and phloem-mobile:** Xylem

- **Site of initial deficiency symptoms:** Structural parts of the plant, such as stocks and stems. Plants may appear soft and wimpy. Deformity in the seed heads.
- **Role in the plant:** Metabolic catalyst, photosynthesis, reproduction, improving flavor and sugar content, impacting nitrogen utilization, lignification.
- **Toxicity:** Copper toxicity will generally show up as chlorosis or reddish-brown necrotic lesions. Toxic levels of copper in the soil will block the uptake of other metals, giving the appearance of manganese, iron, or zinc deficiencies in the top growth.

Zinc

Zinc is another trace element that tests over 90 percent of the time below the 0.02 ppm lab detection limit on the paste test. Excessive phosphorus and high pH levels quickly render zinc unavailable. As with copper, a chelated soil application or foliar application is the best way to overcome the problem. Zinc sulfate can be foliar fed, but soil applications will get tied up rather quickly.

- **Ideal media level:** Standard soil test 20–40 ppm, paste test 0.1–0.2 ppm
- **Mobile in the plant:** No
- **Xylem- and phloem-mobile:** Xylem
- **Site of initial deficiency symptoms:** Zinc deficiencies will first be seen at the new growth of both root and shoot tips. The newest leaves will appear smaller than normal and yellow. Deficient plants will exhibit poor seed production. Overall plant will be short and squat.
- **Role in the plant:** Zinc directly influences chlorophyll production, auxin production, flowering, seed production, and transformation of sugars to starch and is involved in many enzyme systems, such as

> RNA production and protein synthesis. It will help control cell division, especially at the root and shoot tips.
> - **Toxicity:** Toxicity rarely occurs; liming and extra phosphorus applications will fix most high-zinc situations.

Molybdenum

Molybdenum is an anion and therefore less affected by pH or high levels of phosphates. It is required in such small amounts that when trying to adjust the mix with minuscule amounts, it is extremely difficult to get it spread evenly in the mix. Foliar feeding during vegetative stage is the most efficient. Molybdenum will improve nitrogen efficiency, helping to minimize nitrogen applications. It is not a good fix to put more nitrogen on when molybdenum is low. Nitrates will accumulate in the plants, making them more susceptible to disease and insect pressure.

> - **Ideal media level:** Standard soil test 0.04–0.06 ppm
> - **Mobile in the plant:** Poorly
> - **Xylem- and Phloem-mobile:** Yes
> - **Site of initial deficiency symptoms:** New leaves will become narrow and deformed. Whip tail in brassicas is a good example. Chlorosis and necrosis of the newer growth.
> - **Role in the plant:** Critical for the development of the nitrate reductase system, which prevents the accumulation of nitrates. Molybdenum aids in the absorption of atmospheric nitrogen in the root nodules of legumes. It helps convert inorganic phosphorus to an organic form in the plant. Molybdenum helps to increase amino acid levels in the plant.
> - **Toxicity:** Toxicity is very rare, but in studies where toxic levels were given to plants, a yellow-to-orange chlorosis appeared on the leaves. Most plants can tolerate relatively high levels. The toxicity comes

> when using plants high in molybdenum for animal feed. Levels in the plants greater than 5 ppm for sheep and 10 ppm for cattle will cause diarrhea and interfere with copper and selenium utilization.

Silicon

Silicon is almost always deficient in organic mixes due to the absence of soil clay and sand fractions. Silicon shortages can be addressed with the addition of rice hulls as part of the bulking materials. Wollastonite, a mined calcium silicate, can also be used to supply calcium and silicon. It will raise the pH, so if you are adding calcium silicate, do that first, then add the correct type and amount of lime. Calcium silicate is a low-solubility product, so getting a micronized or fine calcium silicate is advisable. The most-available silicon is mono-silicic acid, found in various products that can be used as a soil or foliar application. I prefer to apply the mono-silicic acid as a foliar first; however, it can be applied through drip irrigation as well.

- **Ideal media level:** Standard soil test 40–60 ppm
- **Mobile in the plant:** Poorly
- **Xylem- and phloem-mobile:** Xylem
- **Site of initial deficiency symptoms:** Silicon deficiency is less visual than other nutrients. Deficiencies tend to be more of a breakdown in disease and pest suppression, as well as poor structural integrity.
- **Role in the plant:** Originally, silicon was thought to impart a protective coating around the plant. Now we know that it's much deeper than that. It does suppress disease, as well as reduce some insect pressure. Silicon is a real stress reliever, improving water efficiency and salt tolerance. It also improves structural integrity of the roots, stems, and leaves. There is also an improvement in shelf life for produce.
- **Toxicity:** No known toxicity issues

Table 4: Partial list of nitrogen sources

Product	Type	Effect on pH	% N	% P$_2$O$_5$	% K$_2$O
Urea	Pellet	Acidic	46	0	0
Aqua ammonia	Liquid	Acidic	28–32	0	0
Ammonium nitrate	Granular	Acidic	33	0	0
Ammonium sulfate	Granular	Acidic	21	0	0
Diammonium phosphate	Pellet	Acidic	18	49	0
Monoammonium phosphate	Pellet	Acidic	11	52	0
Calcium nitrate	Pellet	Alkaline	15	0	0
Sodium nitrate	Granular	Alkaline	16	0	0
Potassium nitrate	Granular	Alkaline	14	0	37
Fish meal	Meal	Acidic	10	8.2	0.5
Blood meal	Meal	Acidic	13	0.6	0.4
Corn gluten meal	Meal	Acidic	7	1	0.3
Feather meal	Meal	Acidic	13	1.4	1.8
Crab meal	Meal	Acidic	5	3.6	0.5
Alfalfa meal	Meal	Neutral	3	0.7	2.9

Table 5: Nutrient amounts in selected animal manures

Product	% DM*	lb N/ton	lb P$_2$O$_5$/ton	lb K$_2$O/ton
Dairy manure	18	9	4	10
Horse manure	46	14	4	14
Chicken manure w/o litter	45	33	48	34
Chicken manure w/ litter	75	56	45	34
Turkey manure w/o litter	29	20	16	13
Swine manure w/o bedding	10	10	9	3
Beef manure, feedlot	15	11	7	10
*Dry matter				

Table 6: Partial list of sulfur sources

Product	Type	Effect on pH	% S
Elemental sulfur	Granular	Acidic	90
Ammonium thiosulfate	Liquid	Acidic	26
Ammonium sulfate 21-0-0-24	Granular	Acidic	24
Potassium magnesium sulfate (K-Mag)	Granular	Neutral	22
Sulfate 0-0-0-50	Granular	Neutral	18
Magnesium sulfate (Epsom salts)	Granular	Neutral	13
Calcium sulfate (gypsum)	Powder	Neutral	12

Table 7: Partial list of phosphorus sources

Product	Type	% P_2O_5	% P
Monoammonium phosphate	Granular	52	22.7
Diammonium phosphate	Granular	46	20.1
Liquid polyphosphate 10-34-0	Liquid	34	14.8
Rock phosphate	Dust	25	10.9
Bone meal	Meal	22	9.6
Fish meal	Meal	8.2	3.6
Crab meal	Meal	3.6	1.6
Chicken manure w/o litter, 45% DM		2.4	1.1
Chicken manure w/ litter, 75% DM		2.2	1.0
Turkey manure w/o litter, 29% DM		0.8	0.35
Swine manure, no bedding, 18% DM		0.5	0.2
Beef manure on concrete, 15% DM		0.4	0.2
Seabird guano		10	4.4

Table 8: Partial list of calcium sources

Product	% Ca	Relative neutralizing value*
Calcitic limestone	30–35	85–100
Dolomitic limestone	20–25	95–110
Gypsum	23	0
Burnt lime (calcium oxide)	55–65	150–180
Rock phosphate	33	
Bone meal	22–26	
Calcium nitrate	20	
Wood ash	25–45	
Oyster shell flour	33–38	variable particle size†
Aragonite (marine base)	34–36	with trace elements†
Poultry layer manure	3–4	
Wollastonite (calcium silicate)	26	low to fair solubility

*Compared to pure calcitic limestone (calcium carbonate)
†Oyster shell or aragonite will be picked up by the standard soil test, but may not show up quickly on the paste test due to their low solubility

Table 9: Partial list of magnesium sources

Product	% Mg
Magnesium sulfate (Epsom salts)	17
Dolomitic lime	11
K-Mag (potassium magnesium sulfate)	11
Magnesium oxide	54
Magnesium oxysulfate	36
Wollastonite (calcium silicate)	6–7
Wood ash	3–5

Table 10: Partial list of potassium sources

Product	% K₂O*	% K
Potassium chloride 0-0-60	60	50
Potassium carbonate liquid 0-0-30	30	24.9
Potassium acetate liquid 0-0-20	20	16.6
Potassium sulfate 0-0-50	50	42
Potassium nitrate 14-0-44	44	36
K-Mag (potassium magnesium sulfate)	22	19
Kelp meal	10	8
Alfalfa meal	2.4	2
Chicken manure w/o litter, 45% DM	1.7	1.4
Chicken manure w/ litter, 75% DM	1.7	1.4
Turkey manure w/o litter, 29% DM	0.7	0.6
Swine manure, no bedding, 18% DM	0.2	0.17
Beef manure on concrete, 15% DM	0.5	0.4
Horse manure, 46% DM	0.7	0.6
Greensand (glauconite)	5	4
Wood ash	3.5–8.5	3.7
K*ash (sunflower hull ash)	34	28.2

* When you buy potassium, the percentage on the bag is K_2O equivalent, not actual potassium. Soil and paste analyses report the potassium as actual K.

Table 11: Partial list of trace element sources

Product	% element
Borax (sodium borate), dry	10–14
Solubor (sodium borate), dry concentrate	20.5
Ferrous sulfate	19–21
Ferrous oxide	77
Chelated iron	5–14
Manganese oxide	63
Manganese carbonate	31
Manganese sulfate	27
Chelated manganese	10–12
Copper sulfate	25
Copper oxide	75–89
Chelated copper	9–13
Zinc sulfate	36

Zinc oxide	78
Zinc carbonate	52
Chelated zinc	9–15
Ammonium molybdate	54
Sodium molybdate (dry)	39
Sodium molybdate (liquid)	10
Calcium silicate (mined)	24
Aragonite (sea-based)	55
Magnesium silicate	28
Potassium silicate	18
Mono-silicic acid (rice hulls)*	1.5
Mono-silicic acid (Sil-Force)†	2.5
Azomite (volcanic rock dust)	variable
* CropSIL product from Nuvia Technologies †Sil-Force product from Nutech	

Chapter 4:

Water and Its Effect on Growing Media

Media with water.
(Grow Ohio)

I consider there to be four basic water sources for irrigation. Going from best to worst, my favorite is reverse osmosis (RO) water. Rainwater is a very close second, with surface water from lakes or rivers coming in third and well water being last. To be fair to well water, it can be very good, but it can also be very bad. The deeper the well, the worse it tends to be. Table 12 shows the differences between municipal, rain, and well water sampled locally in northwest Ohio. I did not include RO water since there is practically nothing in it.

My rating of water sources is solely based on the level of soluble salts in the water. When it comes to cost, the order is nearly reversed. If you want very clean water, reverse osmosis water will provide you that, but at a higher cost. It will also increase the amount of water needed in the system, because the backwash systems waste a lot of water to keep the filters clean. Rainwater would be my water source of choice based on quality and cost, but I get it, some locations don't get enough rain and storing water may not be an option. A one-inch rainfall provides roughly 27,000 gallons per acre.

Municipal water usually comes from one of two types of sources—surface water from lakes or rivers or well water. Smaller municipalities may pull their water from a well, but most larger towns depend on surface water sources. Water quality from either a well or surface water source can vary considerably. If you can source water directly from a river, the quality will probably be related to how close you are to the headwater source. Well water will be all over the map. *The bottom line is to test your water before starting a growing operation.* I recently got a call from a grower who had seven hoop houses already built when he got the water tested, only to find out that the soluble salts level was around 750 ppm. Surface water sourced directly from a river should be tested once in the spring during high flow and again in late summer or early fall during low flow. Municipalities who pull from a river and hold it in reservoirs will tend to even out the possible variations from spring to fall, requiring only one water test. Lake water sources tend to be very good, but test just in case.

Various Water Sources

Table 12: Soluble salts in selected water sources in northwestern Ohio. Analysis by Logan Labs.

Parameters	Rainwater	Municipal water	Well water
pH	5.9	7.2	6.8
Conductivity (dS/m)	0.02	0.68	1.82
Sodium adsorption ratio	0.02	1.25	0.58
Calcium (ppm)	1.5	73.8	289.8
Magnesium (ppm)	0.2	13.1	48.7
Potassium (ppm)	0.3	5.1	2.4
Sodium (ppm)	0.1	44.5	40.5
Total alkalinity (ppm)	5.5	29.0	101.5
Bicarbonate (ppm)	7.0	35.0	124.0
Chloride (ppm)	2.0	139.0	50.0
Sulfate (ppm)	0.9	171.5	905.7
Salt concentration (ppm)	14.7	435.8	1164.2
Boron (ppm)	<0.02	0.12	0.42

Growing in media is generally done under a cover, such as a hoop house, or in a permanent structure. This results in minimal leaching capability, so the cleaner the water, the better. However, before you breathe a sigh of relief because you have RO or rainwater, there is another set of issues to deal with when using ultra-clean water.

Rainwater or RO water are considered stripping waters. They are so clean that the soluble nutrients in your mix can easily dissolve into the irrigation water and move deeper into the soil bed. No problem, right? The deep roots will pick up the necessary nutrients down below and move them up into the plant. For the first few growing cycles that may be true, but over time the excess nutrients become a soluble salts issue. The nutrients that move by mass flow will be the quickest to migrate deeper into the bed. Nutrients such as calcium, magnesium, nitrate, boron, and molybdenum will be the first to migrate. Even nutrients such as potassium, which moves by diffusion and is held on the exchange sites of clays in a normal soil, will easily move down because of the lack of clays and exchange sites in growing media. Phosphorus will normally not move since it precipitates out with a variety of cations. If you test your media to a depth of 6 inches and make amendments based on that shallow test, you will be continually building up the soluble salts in the lower levels of your bed. Once you get an electrical conductivity (EC) reading into the 800–1,000 ppm range, the roots will start to shallow up, increasing the plants' demand for water and exacerbating the problem. Testing your beds after every grow cycle is a great idea, but be sure to sample the entire depth of the bed, preferably in split depths. For example, if your bed is 12 inches deep, take a sample from 0–6 and 6–12 inches and analyze separately. You are not going to stop this downward migration of nutrients, therefore sample the entire bed so you are not neglecting valuable nutrients down below. Plants growing in media will root down very quickly, picking up mobile nutrients. Calcium solubility will need to be maintained at the surface since roots need calcium at the growing points to be able to move in the soil.

Starting out with a fairly hot/salty medium (>800 ppm) will quickly start to limit deep rooting. The deeper the bed, the lower the overall nutrient base needs to be (see Table 3). I start to see salt burn on plants when soluble salt levels reach 800–1,000 ppm at a 6-inch depth. Taking some additional samples at depths lower than 6 inches generally returns soluble salt values of 1,500–2,000 ppm or higher. When this occurs, a complete makeover of the bed is necessary. Unfortunately, it is the lower part of the bed that needs to be removed and replaced with new bulking material. Starting out with lower nutrient levels delays this problem, and it is always easier to add nutrients to the bed than to try to remove minerals from the existing mix. When you start to detect a buildup of salts in the lower level, occasionally watering to the point of runoff at the bottom of the bed will help flush excess salts. This is best done at the end of a grow cycle and before retesting and replanting.

In order to accomplish a lower but adequate nutritional mix, make sure to run a media density test on your bulking mix. You can use the media density to compare your bulking mix weight with the weight of a normal soil. Most growing media range in weight between 20–40 percent of a normal soil. Therefore, any amendments that you are going to add to your mix should be reduced by the media density. For example if your media had a media density of 0.25 g/cm^3 and we assume that a normal soil has a bulk density of 1.0 g/cm^3, all your nutrient amendments should be 25 percent of what you would add to a normal soil. The fact that growing media don't have clays and exchange sites to hold nutrients in place means that you don't need normal soil fertility levels. Instead, the nutrients in growing mixes are free-floating, highly mobile, and flow with the water. Most of the nutrients can be picked up deeper in the bed because of the increased root mass that develops in media. These calculations will be dealt with in more detail in the balancing section.

What about growing in pots? Large pots—30 gallons or larger—should be treated like growing beds. It comes down to the amount of soil per plant. The larger the amount of soil used in growing a plant, the lower the overall fertility needs to be. A certain level of nutrients is required, depending on the type and size of the plant. As plants grow, the demand for certain nutrients increases and these requirements must be met or deficiencies will occur. When growing transplants, the nutrient requirement is very low; however, when taking a plant to full maturity, all this changes and so does the level of salt tolerance. The lower the amount of soil used to take a plant to maturity, the more difficult fertilization becomes. It becomes a constant battle between deficiencies and salt toxicity. Watering regimes are shorter, further increasing the need for top-quality water. A water-only mix for containers less than 10 gallons and for sure less than 5 gallons is almost impossible to put together without having soluble salt issues. Small containers are going to require frequent watering, along with constant fertilization and a huge amount of monitoring and management. It is always important to be able to leach the pots or beds, but it is especially important in small pots since soluble salts can easily get out of hand. High plant densities that provide a high return on investment may look very good on paper, but the monitoring and management had better be top notch.

Chapter 5:

Nutrient Balancing in Media

Nutrient balancing.
(Grow Ohio)

This chapter will deal with balancing a bulking mix in order to end up with a well-balanced growing medium. This will be set up primarily as a water-only mix for at least one cycle or more, depending on the crop. If you are growing leafy greens, you might get by for three or four cycles by only adding additional nitrogen. Running a paste test after a couple of cycles is always a good idea. Nitrogen levels would be my main concern, so when ordering a paste test, be sure to request available nitrogen. This will not be a normal potassium chloride extract like is done on a standard soil test; it will be a nitrogen analysis of the water extract itself. Most of the protein nitrogen sources will be converted into nitrate nitrogen after the first cycle, so analyzing the water extract will be a pretty good indicator of the level of available nitrogen in the growing mix.

Balancing soluble salts

The bulking mix analyzed in Figures 7A and 7B is a blend of manure, bark compost, leaf compost, peat, and rice hulls. The standard soil test (Figure 7A) is very good, based on general levels. The pH is at a good level and the overall fertility levels are not excessive, making it a good candidate for a bulking agent or even a growing medium with some minor adjustments. The paste test (Figure 7B) suggests the need for some amendments before becoming a growing medium. As it is, it would be a great bulking mix to dilute some existing mixes that have excessive nutrients.

The growing medium analyzed in Figures 8A and 8B is an example of a medium with way too much nutrition, resulting in a high soluble salts problem. This sample is from a cannabis facility which has been over-fertilized for quite some time. This is the number one problem that I see when consulting with various clients growing in media. There are basically two choices to fix this problem, not including dumping the mix and starting over. You can either try and flush the mix with lots of good water or try to dilute the mix with a bulking agent. Dumping the mix is expensive and flushing the mix is messy and very difficult to know when to stop flushing. I feel the best bet is to dilute the mix with a bulking material. With the proper data, you can calculate very accurately how to best dilute the problem mix. Let's delve into the best way to accomplish a good dilution.

The standard test in Figure 8A indicates that the sample was collected at a 6-inch depth. If this is the total depth of the bed, all we need to do is dilute the soluble salts down to 500–600 ppm, or by roughly 50 percent. However, if the bed is 12 or 18 inches deep, the soluble salts in the lower levels are more than likely to be much higher than the surface

Sample Location			New Compst
Sample ID			Blend
Lab Number			11
Sample Depth in inches			9
Total Exchange Capacity (M. E.)			4.57
pH of Soil Sample			6.6
Organic Matter, Percent			>20
ANIONS	SULFUR:	p.p.m.	16
	Mehlich III Phosphorous:	as (P_2O_5) lbs / acre	220
EXCHANGEABLE CATIONS	CALCIUM: lbs / acre	Desired Value	
		Value Found	1386
		Deficit	
	MAGNESIUM: lbs / acre	Desired Value	
		Value Found	358
		Deficit	
	POTASSIUM: lbs / acre	Desired Value	
		Value Found	737
		Deficit	
	SODIUM:	lbs / acre	95
BASE SATURATION %	Calcium (60 to 70%)		50.60
	Magnesium (10 to 20%)		21.78
	Potassium (2 to 5%)		13.80
	Sodium (.5 to 3%)		3.00
	Other Bases (Variable)		4.80
	Exchangable Hydrogen (10 to 15%)		6.00
TRACE ELEMENTS	Boron (p.p.m.)		0.57
	Iron (p.p.m.)		25
	Manganese (p.p.m.)		24
	Copper (p.p.m.)		< 0.2
	Zinc (p.p.m.)		2.19
	Aluminum (p.p.m.)		13
OTHER	Cobalt ppm		< 0.02
	Molybdenum ppm		0.04
	Ammonium (p.p.m.)		0.3
	Nitrate (p.p.m.)		2.9
	Selenium ppm		< 0.02
	Silicon ppm		10.4
	Estimated Nitrogen Release #N/Acre		N/A
	EC mmhos/cm		0.28
	Media Density g/cm3		< 0.25

Figure 7A: *Standard soil test results for bulking mix*

Sample Location			New Compst
Sample ID			Blend
Lab Number			208400
Water Used			DI
pH			6.6
Soluble Salts		ppm	263
Chloride (Cl)		ppm	80
Bicarbonate (HCO3)		ppm	112
ANIONS	SULFUR	ppm	1.87
	PHOSPHORUS	ppm	9.26
SOLUBLE CATIONS	CALCIUM	ppm	17.09
		meq/l	0.85
	MAGNESIUM	ppm	14.90
		meq/l	1.24
	POTASSIUM:	ppm	72.91
		meq/l	1.89
	SODIUM	ppm	3.00
		meq/l	0.13
PERCENT	Calcium		20.74
	Magnesium		30.13
	Potassium		45.96
	Sodium		3.17
TRACE ELEMENTS	Boron (p.p.m.)		0.4
	Iron (p.p.m.)		0.73
	Manganese (p.p.m.)		1.4
	Copper (p.p.m.)		0.08
	Zinc (p.p.m.)		0.26
	Aluminum (p.p.m.)		0.5
OTHER			

Figure 7B: *Paste test results for bulking mix*

Sample Location	Problem
Sample ID	Bed
Lab Number	11
Sample Depth in inches	6
Total Exchange Capacity (M. E.)	17.33
pH of Soil Sample	6.8
Organic Matter, Percent	>20

ANIONS			
	SULFUR:	p.p.m.	209
	Mehlich III Phosphorous:	as (P_2O_5) lbs / acre	1378

EXCHANGEABLE CATIONS			
	CALCIUM: lbs / acre	Desired Value	
		Value Found	4636
		Deficit	
	MAGNESIUM: lbs / acre	Desired Value	
		Value Found	607
		Deficit	
	POTASSIUM: lbs / acre	Desired Value	
		Value Found	1186
		Deficit	
	SODIUM:	lbs / acre	174

BASE SATURATION %		
	Calcium (60 to 70%)	66.86
	Magnesium (10 to 20%)	14.59
	Potassium (2 to 5%)	8.77
	Sodium (.5 to 3%)	2.19
	Other Bases (Variable)	4.60
	Exchangable Hydrogen (10 to 15%)	3.00

TRACE ELEMENTS		
	Boron (p.p.m.)	0.75
	Iron (p.p.m.)	181
	Manganese (p.p.m.)	16
	Copper (p.p.m.)	2
	Zinc (p.p.m.)	5.95
	Aluminum (p.p.m.)	133

OTHER		
	Cobalt ppm	0.078
	Molybdenum ppm	0.03
	Ammonium (p.p.m.)	0.3
	Nitrate (p.p.m.)	158.7
	Selenium ppm	0.47
	Silicon ppm	11.5
	Estimated Nitrogen Release #N/Acre	N/A
	EC mmhos/cm	1.14
	Media Density g/cm3	0.28

Figure 8A: *Standard soil test results for problem mix*

Sample Location			Problem
Sample ID			Bed
Lab Number			227051
Water Used			DI
pH			6.8
Soluble Salts		ppm	1,106
Chloride (Cl)		ppm	111
Bicarbonate (HCO3)		ppm	40
ANIONS	SULFUR	ppm	148.8
	PHOSPHORUS	ppm	7.2
SOLUBLE CATIONS	CALCIUM	ppm	140.30
		meq/l	7.02
	MAGNESIUM	ppm	59.95
		meq/l	5.00
	POTASSIUM:	ppm	118.40
		meq/l	3.08
	SODIUM	ppm	50.21
		meq/l	2.18
PERCENT	Calcium		40.62
	Magnesium		28.93
	Potassium		17.81
	Sodium		12.64
TRACE ELEMENTS	Boron (p.p.m.)		0.13
	Iron (p.p.m.)		1.41
	Manganese (p.p.m.)		0.12
	Copper (p.p.m.)		< 0.02
	Zinc (p.p.m.)		0.03
	Aluminum (p.p.m.)		1.33
OTHER			

Figure 8B: *Paste test results for problem mix*

1106 ppm. Using an electrical conductivity (EC) meter and testing the lower levels will verify if this is true or not. Remember that the paste soluble salts readings are taken after holding the media in water/paste condition for 24 hours, so follow the same protocol for your samples. Take your samples at 6-inch increments starting at the top and run your own analysis. Calibrate your EC meter by comparing your EC measurement of the top 6-inch sample with that of the lab's paste analysis. Samples held overnight in the paste condition almost always have a higher EC reading than the standard soil test procedure.

If the lower depths of the bed have a higher EC than the surface, you should not include them in the dilution process. The depth that you sampled for the lab analysis should be what you use for your dilution process. This cannot be done in the bed. That layer should be taken out and put on plastic or another clean surface for mixing. The remaining material in the bed does not need to be disposed of. Pile outside and let the rainwater flush it out over time. Leaves or plant-based compost could be incorporated as a dilution material. Diluting the problem mix with an equal amount of the bulking material will reduce the soluble salts to a value halfway between the levels of the problem mix and the bulking material.

For example, the problem mix shown in Figure 8B has a soluble salts level of 1106 ppm. The bulking mix we are using to dilute it has a soluble salts level of 263 ppm. To calculate the soluble salts level of the new mix made from equal parts of problem mix and bulking mix, we add together their soluble salts levels (1106 ppm + 263 ppm = 1369 ppm). Then we divide the sum by two to get the soluble salts level of the new mix (1369 ppm ÷ 2 = 684.5 ppm). Or, if we are mixing different proportions of the bulking mix and the problem mix, we can calculate the new soluble salts level using the following equation:

Equation 1: Soluble salts level of new mix

$$(B_{ss} \times B_{\%}) + (P_{ss} \times P_{\%}) = N_{ss}$$

B_{ss} = Soluble salts level of bulking media, in ppm
$B_{\%}$ = Percent of bulking media in new mix, in decimal form
P_{ss} = Soluble salts level of problem mix, in ppm
$P_{\%}$ = Percent of problem mix in new mix, in decimal form
N_{ss} = Soluble salts level of new mix, in ppm

If the bed is only 6 inches deep, you might stop with a ppm of 684.5 as long as you are not germinating seeds in the mix. Transplants placed into the bed will be more tolerate of the salt level than germinating seeds. For a bed deeper than 6 inches, I would prefer to push the soluble salts down to around 500 ppm in order to slow the deep accumulation of salts over time. You can calculate exactly how much bulking mix to add to the problem mix to reach your desired soluble salts concentration using the following equation:

Equation 2: Amount of bulking mix to add to reach desired soluble salts concentration

$$B_{\%} = \frac{P_{ss} - N_{ss}}{P_{ss} - B_{ss}}$$

Example 1: The problem mix analyzed in Figure 8B has a soluble salts level of **1106 ppm**. What percentage of bulking material (soluble salts level of **263 ppm**) should we add to the mix to get a new mix with a soluble salts level of **500 ppm**?

Solution 1: First, we need to figure out which numbers to plug into Equation 2. P_{ss}, the soluble salts level of the problem mix, is **1106 ppm**. N_{ss}, the soluble salts level of the desired new mix, is 500 ppm. B_{ss}, the soluble salts level of the bulking mix, is **263 ppm**. Next, we plug those numbers into **Equation 2:**

$$B_\% = \frac{1106 \text{ ppm} - 500 \text{ ppm}}{1106 \text{ ppm} - 263 \text{ ppm}}$$

Then we follow the standard order of operations by doing the subtraction first:

$$B_\% = \frac{606 \text{ ppm}}{843 \text{ ppm}}$$

Finally, we divide the numerator by the denominator to get a solution of **0.72**, or 72 percent. This means that, for the new mix to have a soluble salts level of 500 ppm, we will need to mix 72 percent of the bulking mix with 28 percent of the problem mix.

You can always round the results up or down a little to make measuring the mixes easier. In this case, 70 percent bulking mix and 30 percent problem mix will give a new soluble salts level of 515.9 ppm, which will be close enough. I think 70 percent is a good place to stop but you could continue to dilute, although if you go much further, you might as well remove all the mix and go with the bulking material and rebalance. Using a straight sphagnum peat with a lower soluble salts content would achieve a quicker dilution using more of your original mix.

Just don't assume that the whole bed has the same soluble salts content as the top of the 6-inch sampled bed. I am speaking from experience and it is more than likely that if you make that assumption, you will go through a lot of work diluting the mix only to find that the final product is nowhere near as diluted as you think it should be. Once you decide on the percent dilution, and as long as you have a complete test on the bulking material and your problem bed, you can use Equation 1 for your nutrients and see if you still have adequate levels of nutrients for production.

Example 2: You've decided to dilute the problem mix from Figure 8B (**140.3 ppm** Ca) with **70 percent** of the bulking mix (**17.1 ppm** Ca). How much soluble calcium will be in the new mix?

Solution 2: This time, we will use **Equation 1**, except instead of putting in the soluble salts values, we will put in the calcium values instead:

$$(B_{Ca} \times B_{\%}) + (P_{Ca} \times P_{\%}) = N_{Ca}$$

$$(17.1 \text{ ppm} \times 0.7) + (140.3 \text{ ppm} \times 0.3) = N_{Ca}$$

We do the multiplication in parenthesis first:

$$(12.0 \text{ ppm}) + (42.1 \text{ ppm}) = N_{Ca}$$

Then we add the two numbers to get a solution of **54.1 ppm** soluble calcium in the new mix.

This level of calcium would be fine for most greenhouse crops, but plants like cannabis would probably do better with another 50 ppm of calcium. Do these calculations for all the nutrients, and you should get in the ballpark of where the blended mix will be. If you don't have a complete test on the bulking material but you know the EC, you can still work out the necessary dilution percentage and do the dilution. Once completed, then send in a sample to the lab and check the available nutrient levels with a complete test.

Rebalancing the New Blend mix

In order to rebalance the New Compost Blend mix analyzed in Figures 7A and 7B, it is important to understand that the soil test results are reported in pounds per acre and that we need to scale it down from pounds per acre in a 6-inch furrow slice to something more reasonable for growing in beds or pots, like pounds or grams per cubic foot. One critical part of this process is to use the media density to compensate for the weight difference between normal field soils and a mostly organic, lightweight mix. This will help prevent over-fertilization and excessive salt issues. This also means that when making amendments we get away from the idea of adding a cup of this or that to the media. It must be done in weight and not volume. As long as you know the weight of the material in the cup, that's okay. Many of the meals such as alfalfa meal, feather meal, etc. used in media mixes will weigh 35–45 pounds per cubic foot, whereas mineral amendments will weigh up to 60 pounds per cubic foot, so a cup of each is not equal.

The soil reports in Figures 7A and 7B show that the New Blend bulking mix has a fairly low total exchange capacity, indicating minimal amounts of cations in the mix. Conversely, had the total exchange capacity been high (>16 meq/100 g), that would have indicated too many cations, making it a bad bulking mix with a potentially high level of soluble salts. The standard soil EC reading measured in dS/m can be converted (with an accuracy of ± 10 percent) to ppm by multiplying it by 640. In this case, an EC of 0.28 dS/m translates into 179.2 ppm of soluble salts, which is low, making it a good bulking agent. The standard EC reading will almost always be lower than the paste soluble salts reading, primarily due to the holding time before taking a reading. The paste test is read after nearly 24 hours, whereas the standard soil test EC is read after several minutes.

Since the New Blend mix has a low EC, that is a good indication that amendments will have to be added if it is going to be used as growing media mix. I make most of my amendment additions based on the paste test since it is these nutrients that the plants will see. The paste test indicates that the sulfur, calcium, and magnesium are low, and the standard nitrogen test shows a very low level.

Calculating the amount of fertilizer to add for each nutrient based on the soil test results is slightly more complicated for growing media than for normal soils. For a normal soil that had the test results shown in Figure 7A, we would assume that every three inches' depth of soil weighs approximately a million pounds per acre. Since this sample was taken at a 9-inch depth, that makes the total weight of the soil 3 million pounds per acre. So to raise a nutrient level by 1 ppm, we need to add 3 pounds per acre of that nutrient to the soil. If we know the concentration of the target nutrient in our desired amending product, we can calculate the amount of product to apply using the following equation:

Equation 3: Amount of product to apply, in pounds per acre

$$N = \frac{DP}{3C}$$

D = Depth of sample analyzed, in inches
P = Parts per million of the nutrient to add to the mix
C = Concentration of the nutrient in the amending product, in decimal form
N = Amount of amendment to add, in pounds per acre

Example 3: To balance the New Blend mix for vegetable production, we need to increase the calcium level from **17.1 ppm** to **50.0 ppm**. If we are using gypsum as a calcium source, how many pounds per acre do we need to add?

Solution 3: First, we need to get our numbers to plug into Equation 3. Looking at Figure 7A, we see that the depth of the sample analyzed is 9 inches. To figure out how much calcium we need to add to the mix, we subtract **17.1 ppm** from **50.0 ppm** to get **32.9 ppm**. To figure out the concentration of calcium in gypsum, we look back at Table 8 and see that gypsum is 23 percent calcium (**0.23** in decimal form). Now we can plug those values into Equation 3:

$$N = \frac{DP}{3C}$$

$$N = \frac{(9)(32.9)}{3(0.23)}$$

Following standard order of operations, we do the multiplication first:

$$N = \frac{296}{0.69}$$

Finally, we divide the numerator by the denominator to get a solution of **430**. So we need to apply 430 pounds of gypsum per acre to a field to raise the calcium concentration by the desired amount.

Unfortunately, the standard pounds-per-acre figure is not very useful when we are working with growing media, for two reasons. First of all, most of the time we are not talking about adjusting acres, but hundreds or a few thousand square feet. So let's shrink things down to a thousand square feet, such as a garden, and then even smaller for beds or pots. There are 43,560 square feet in an acre, or 43.56 thousand-square-foot parcels in an acre. So, to calculate the amount of amendment needed per thousand square feet, you can just divide the result of Equation 3 by 43.56, which will change the 3 in the denominator to 131 (See Equation 4, next page).

However, if we are using raised beds or pots, we are not dealing with square foot area, but rather cubic feet. To calculate the number of cubic feet, just multiply the surface area by the depth in feet. You can convert the depth in inches to feet by dividing the depth in inches by 12. For example, to convert 9 inches to feet, divide 9 by 12 to get 0.75 feet. For a 1,000 ft^2 bed that is 9 inches deep, just multiply 1,000 by 0.75 to get 750 ft^3.

Second, we need to account for the media density. The media density of the mix in Figure 7A is only 0.25 g/cm^3. Normal mineral soils usually have a bulk density ranging from 0.9–1.7 g/cm^3. To make the math simpler, we will assume that a "normal" soil has a bulk density of 1.0 g/cm^3. That means that we can just treat the media density as a percentage of the bulk density of a normal soil in our nutrient calculations. So we can just multiply our pounds-per-acre, pounds-per-square foot, or pounds-per-cubic foot result by the media density to get an accurate amount of product to apply.

Finally, if we are dealing with smaller containers or beds, the result in pounds will be a very small number. So we will also want to convert pounds to grams by multiplying by 454.

This seems like a long process to come up with the necessary amount, but after doing it a couple of times it will be much easier. Here is a simplified equation for calculating the amount of product needed in pounds per 1,000 ft²:

Equation 4: Amount of product to apply, in pounds per thousand square feet

$$N = \frac{DPM}{131C}$$

D = Depth of sample analyzed, in inches
P = Parts per million of the nutrient to add to the mix
M = Media density, in g/cm³
C = Concentration of the nutrient in the amending product, in decimal form
N = Amount of amendment to add, in pounds per 1,000 ft²

And here is the simplified equation for calculating the amount of product needed in grams per cubic foot. This equation is derived by using a value of 12 inches for D in Equation 4 (which eliminates depth from the final equation because the results are in volume, not surface area), dividing the result by 1000 to get cubic feet, and then multiplying by 454 to get a solution in grams.

Equation 5: Amount of product to apply, in grams per cubic foot

$$N = \frac{PM}{24C}$$

P = Parts per million of the nutrient to add to the mix
M = Media density, in g/cm³
C = Concentration of the nutrient in the amending product, in decimal form
N = Amount of amendment to add, in g/ft³

Example 4: To balance the New Blend mix for vegetable production, we need to increase the calcium level from **17.1 ppm** to **50.0 ppm**. If we are using gypsum as a calcium source, how many grams do we need to add to a bed that is **4 feet long, 4 feet wide, and 18 inches deep?**

Solution 4: Since we want to know how many grams of gypsum to apply and this is a small bed, we will use **Equation 5**. First, we need to figure out what values to plug into the equation. Since we want to raise the calcium level from 17.1 ppm to 50.0 ppm, we need to add **32.9 ppm** of calcium. Looking at Figure 7A, we see that the media density is **0.25 g/cm³**. The concentration of calcium in gypsum, as we see from Table 8, is **0.23**.

$$N = \frac{(32.9)(0.25)}{(24)(0.23)}$$

Following basic order of operations, we do the multiplication first.

$$N = \frac{8.23}{5.52}$$

Then we divide the numerator by the denominator to get a solution of **1.49 grams** of gypsum per cubic foot of soil. But we're not quite done—we need to calculate how many cubic feet of soil are in our bed. Since it's 4 feet by 4 feet by 18 inches, we divide 18 inches by 12 to get 1.5 feet and then multiply 4 by 4 by 1.5 to get **24 cubic** feet. Finally, we multiply **1.49 g/ft³ by 24 ft³** to get a solution of **36 grams** of gypsum. That tells us that it will only take 36 grams of gypsum (distributed evenly) to increase the soluble calcium in the bed by 33 ppm, assuming that all the gypsum is soluble and that there is no tie-up of the calcium.

These equations (Equation 4 for larger plots where products are applied in pounds per 1,000 ft^2 and Equation 5 for smaller plots and containers where products are applied in grams per cubic foot) should be used with all cations for improving levels in a mix, as well as trying to lower the level in a modified mix. For example, let's say that you made up a test mix and after testing the mix at the lab, you found that in the paste analysis the calcium was 180 ppm instead of the 100 ppm that you were shooting for. Take the excess 80 ppm and calculate back to figure out how much less lime or gypsum you should add to the next batch of the mix to bring the calcium back in line with your ideal levels.

Phosphorus balancing

Phosphorus is so reactive with other cations and trace metals that I lean on the standard soil report more heavily than I do for most other nutrients. I still want the paste test to fall in the range of 2.5–3.5 ppm, but I really want the standard test to also make the range from 350–500 lb P_2O_5/ac. The standard soil test provides the feed rate of nutrients into solution and, with all the interference issues, it is prudent to have adequate amounts in reserve. This is also the case for some of the trace elements, like manganese and copper, which suffer availability issues from pH interference problems. For mixes with high pH values, the desired numbers on the standard test may have to be pushed higher to satisfy the paste numbers. If the paste test shows a low level of phosphorus solubility, but the standard soil test is >350 ppm, I would put a small amount of a soluble P source such as fish meal in the transplant holes to stimulate rapid root development in the soil media. Excellent biology levels will promote phosphorus solubility, so keep those microbes happy.

Nitrogen balancing

I can't say enough about monitoring nitrogen levels in your mix. *It drives the boat, and when it goes short, all sorts of nutrient levels fall apart very quickly.* Even though most of the modified mixes are composed of organic constituents, the nitrogen release is not the same as from the humus portion of a normal soil. I typically don't see much free nitrogen in growing media on the available nitrogen test from the lab, unless manure compost or another source of nitrogen was included in the initial mix. Samples with high levels of nitrogen may also have excessive levels of phosphorus and potassium if too much manure-based amendment was used. These samples are also the ones that have high salt issues. Table 13 shows some starting nitrogen levels from the available nitrogen lab test for variable crops at the beginning of the growing season.

Table 13: Starting Nitrogen Levels for Various Crops at a 6" Sampling Depth

Crop	Nitrate (ppm)	Ammonium (ppm)
Leafy greens	25–35	<5
Brassicas	40–50	<5
Tomatoes	60–70	<5
Flowers	30–40	<5
Cannabis, 9 weeks to maturity	50–60	<5
Cannabis, 12–16 weeks to maturity*	70–80	<5
*Outdoor operations with inground beds, after cannabis is beginning to flower		

Assuming that you are starting the season with no manure-based products and nitrogen levels are at background levels (5–7 ppm nitrate and ammonium combined), Table 14 indicates the amount of nitrogen needed on a pound basis.

Table 14: Actual pounds of nitrogen required for various crops

Crop	Nitrogen (lb/ac)	Nitrogen lb/1,000 ft²
Leafy greens	50–70	1.1-1.6
Brassicas	80–100	1.8–2.3
Tomatoes	120–140	2.8–3.2
Flowers	60–80	1.4–1.8
Cannabis, 9 weeks to maturity	100–120	2.3–2.8
Cannabis, 12–16 weeks to maturity*	140–160	3.2–3.7
*Outdoor operations with inground beds, after cannabis is beginning to flower		

Using any organic mulch with a carbon:nitrogen ratio greater than 20:1 will begin to tie up nitrogen and will require additional nitrogen.

Trace element rebalancing

As long as the trace elements meet my desired paste levels, I do not feel that extra levels of trace minerals are needed to cover moderately low levels on the standard soil test, unless the pH is above 7.2 and phosphorus is greater than 500 lb P_2O_5/ac. I do encourage adding some azomite or rock dust products to all mixes as a slow-release, all-encompassing trace element package. This could range from 2–4 pounds per 500 ft³ of mix.

When any of the trace elements drops below my desired levels on the paste test, I broadcast the deficient trace elements using sulfated forms for the metals (Fe, Mn, Zn, or Cu) and a borate product for boron. Zinc and copper are generally deficient; probably close to 90 percent of the time they fall below the detection limit (0.02 ppm) at the lab. Zinc is deficient because of excessive phosphorus and copper is deficient just because of the high level of organics in the mix. Manganese is probably short 50–60 percent of the time, especially if the pH is high. Adding trace elements to the mix is difficult at best, mainly because of the low levels needed. We could be looking at 2–6 grams of a trace element for a 4- by 4-foot bed. There is virtually no way to hand-spread evenly that small of an amount on a 4 x 4 bed, unless it is blended with another product such as feather, blood, or alfalfa meal. The absolute best way to add the trace elements is by dissolving them in hot water, spraying them on the bed, and working them in. Any time you put that small of an amount on with everything else, tie-ups will occur, so adding a foliar feeding program to your trace element strategy is essential.

Chapter 6:

Rebalancing a Problem Mix with Peat

Peat moss.
(Wikimedia)

Rather than using a blended bulking mix that a lot of people might not have access to, let's use the most common product available, sphagnum peat moss, to rebalance the problem mix analyzed in Figures 8A and 8B. Figures 9A and 9B shows the lab results from a Canadian peat.

The Canadian peat has a soluble salts level of 54 ppm, which is great for diluting the overall soluble salts level, but it also means that there is virtually no other nutritional value in the product. Therefore, we may

Sample Location			Worm	Canada
Sample ID			Castings	Peat
Lab Number			14	15
Sample Depth in inches			6	6
Total Exchange Capacity (M. E.)			25.24	14.67
pH of Soil Sample			6.8	4.5
Organic Matter, Percent			>20	>20
ANIONS	SULFUR:	p.p.m.	248	16
	Mehlich III Phosphorous:	as (P_2O_5) lbs / acre	79	24
EXCHANGEABLE CATIONS	CALCIUM: lbs / acre	Value Found	7496	1316
	MAGNESIUM: lbs / acre	Value Found	984	455
	POTASSIUM: lbs / acre	Value Found	250	27
	SODIUM:	lbs / acre	74	69
BASE SATURATION %	Calcium (60 to 70%)		74.25	22.42
	Magnesium (10 to 20%)		16.24	12.92
	Potassium (2 to 5%)		1.27	0.24
	Sodium (.5 to 3%)		0.64	1.02
	Other Bases (Variable)		4.60	8.40
	Exchangable Hydrogen (10 to 15%)		3.00	55.00
TRACE ELEMENTS	Boron (p.p.m.)		0.97	0.81
	Iron (p.p.m.)		225	190
	Manganese (p.p.m.)		8	5
	Copper (p.p.m.)		0.87	< 0.2
	Zinc (p.p.m.)		7.15	0.6
	Aluminum (p.p.m.)		48	51
OTHER	Cobalt ppm		0.031	0.027
	Molybdenum ppm		0.02	0.06
	Ammonium (p.p.m.)		0.6	2.6
	Nitrate (p.p.m.)		242.3	1
	Selenium ppm		0.55	0.3
	Silicon ppm		25.7	7.2
	Estimated Nitrogen Release #N/Acre		N/A	N/A
	EC mmhos/cm		1.67	0.08
	Media Density g/cm3		0.47	0.12

Figure 9A: *Standard soil test results for Canadian peat*

Sample Location			Worm	Canada
Sample ID			Castings	Peat
Lab Number			227498	227499
Water Used			DI	DI
pH			6.8	4.5
Soluble Salts		ppm	1,359	54
Chloride (Cl)		ppm	133	31
Bicarbonate (HCO3)		ppm	67	34
ANIONS	SULFUR	ppm	176.4	3.51
	PHOSPHORUS	ppm	0.22	0.32
SOLUBLE CATIONS	CALCIUM	ppm	241.70	3.93
		meq/l	12.09	0.20
	MAGNESIUM	ppm	83.90	2.39
		meq/l	6.99	0.20
	POTASSIUM:	ppm	30.70	1.34
		meq/l	0.80	0.03
	SODIUM	ppm	31.38	9.61
		meq/l	1.36	0.42
PERCENT	Calcium		56.90	23.17
	Magnesium		32.92	23.47
	Potassium		3.75	4.11
	Sodium		6.42	49.25
TRACE ELEMENTS	Boron (p.p.m.)		0.08	0.1
	Iron (p.p.m.)		0.99	0.45
	Manganese (p.p.m.)		0.05	0.04
	Copper (p.p.m.)		< 0.02	< 0.02
	Zinc (p.p.m.)		< 0.02	< 0.02
	Aluminum (p.p.m.)		1.57	0.33
OTHER				

Figure 9B: *Paste test results for Canadian peat*

have to make more amendments to adjust the final product to achieve optimum balance. Using Equation 1, we calculate that diluting the problem mix with **50 percent** Canadian peat will result in a soluble salts level of **580 ppm** in the new mix:

$$(B_{ss} \times B_{\%}) + (P_{ss} \times P_{\%}) = N_{ss}$$

$$(54 \text{ ppm} \times 0.5) + (1106 \text{ ppm} \times 0.5) = 580 \text{ ppm}$$

580 ppm is really a pretty good starting number for a shallow bed, but for deeper beds or pots it might be good to go a little lower, since the level of nutrients in the peat is very low and will probably require additional amendment, which will again increase the soluble salts level. Increasing the level of peat to **60 percent** generates a final soluble salts level of **475 ppm**:

$$(54 \text{ ppm} \times 0.6) + (1106 \text{ ppm} \times 0.4) = 475 \text{ ppm}$$

This may seem incredibly low for some people, but the lower the salts level, the more growing cycles can be completed before major adjustments or a rollover of the bed may be needed. This is especially important for those who are not using rain or RO water. It is easy to make minimal applications of nutrients to the surface of the bed or large pot if the surface gets too low in nutrients, which can be determined through split sampling the bed by depth. Split sampling the bed will really help to determine when or if the bed needs to be completely remixed to minimize high salt levels at the bottom of the bed. Remember, the bigger the root mass, the better the top growth will be, so just sampling the top 6 inches of the media will not provide enough information for the long term.

Table 15: Total nutrient concentrations (ppm) from diluting the problem mix with 60 percent Canadian peat

Nutrient	Problem mix*	Canadian peat[†]	Total
Paste calcium	56.1	2.3	58.4
Paste magnesium	24.0	1.4	25.4
Paste potassium	47.4	0.8	48.2
Paste phosphorus	2.9	0.2	3.1
Paste sulfur	59.5	2.1	61.6
Standard nitrogen[‡]	63.6	2.2	65.8

*Values derived from Figure 8B and multiplied by 0.4 (40 percent)
[†]Values derived from Figure 9B and multiplied by 0.6 (60 percent)
[‡]Sum of nitrate and ammonium values from the standard soil test (8A and 9A)

These numbers would work very well in a general growing mix for vegetables or flowers. For a seed starter mix, I would prefer to use a 70:30 mix, with peat being the bulk of the mix. As is, the 60:40 mix would be a great grow-out mix when using it as a water-only medium with no additional supplements. For a cannabis mix the calcium, magnesium, and potassium should be increased; however, phosphorus and nitrogen are fine, with sulfur being a little high but acceptable.

Table 16: Ideal cannabis nutrient levels (ppm) compared to the current levels in our example mix (Table 15)

Nutrient	Desired	Current	Needed
Paste calcium	110	58.4	51.6
Paste magnesium	35	25.4	9.6
Paste potassium	60	48.2	11.8
Paste phosphorus	2.5	3.1	0
Paste sulfur	12	61.6	0
Standard nitrogen*	60	65.8	0

*For a 6-inch depth

Before we can rebalance the new mix for calcium, we need to determine the pH of the newly formed blend. The pH of the problem blend was 6.8 and the pH of the Canadian peat is 4.5, so the combination of the two will be somewhere between those two numbers. You could easily run a pH test yourself on the new blend by testing a solution of one part media and one part distilled water with a meter or pH paper. In this example, the pH ended up being 5.8. Since we would like to have our mix in the range of 6.2 and 6.5, we need to use a calcium amendment that will raise the pH, so this narrows down the ingredient choice to some kind of lime. But what kind? There are basically two kinds of lime, high-calcium and dolomitic lime. To be fair, calcium silicate could also be used in place of a high-calcium lime, since it will also raise the pH.

Looking at Table 16, I see that I am also going to need both potassium and magnesium, but a lot more calcium. Therefore, a high-calcium type of lime would be my preference to solve the calcium deficiency. Since this is an organic mix and nearly all the organic mixes are deficient in silicon, I would go with a finely ground calcium silicate or Wollastonite. If you can't get Wollastonite or calcium silicate, go with high-calcium lime and foliar feed silicon during the growing season. The potassium and magnesium deficiencies could be resolved using K-Mag, also known as Sul-Po-Mag.

Example 5: How many grams of calcium silicate do we need to add to a **4-foot by 4-foot by 18-inch bed** of our new mix to raise the calcium level by **51.6 ppm**?

Solution 5: Since we want to know how many grams per cubic foot of calcium silicate to apply, we will use **Equation 5**:

$$N = \frac{PM}{24C}$$

P = Parts per million of the nutrient to add to the mix
M = Media density, in g/cm^3
C = Concentration of the nutrient in the amending product, in decimal form
N = Amount of amendment to add, in g/ft^3

First, we need to figure out what numbers to plug into the equation. We know from Table 16 that we want to raise the calcium level by **51.6 ppm**. Figure 8A tells us that the media density of the problem mix is 0.28 g/cm³, and Figure 9A tells us that the media density of the peat is 0.12 g/cm³. We can use **Equation 1** to calculate that the media density of the new mix is **0.18 g/cm³**:

$$(B_{MD} \times B_{\%}) + (P_{MD} \times P_{\%}) = N_{MD}$$

$$(0.12 \text{ g/cm}^3 \times 0.6) + (0.28 \text{ g/cm}^3 \times 0.4) = 0.18 \text{ g/cm}^3$$

By looking at Table 8, we see that Wollastonite (calcium silicate) is **26 percent** calcium. That gives us all the numbers we need to solve Equation 5:

$$N = \frac{(51.6)(0.18)}{(24)(0.26)}$$

$$N = \frac{(9.29)}{(6.24)}$$

Solving Equation 5 gives us a solution of **1.5 g** Wollastonite per cubic foot of mix. Our 4 x 4 x 1.5 ft bed has a total volume of **24 ft³**, so that means that we need to apply **36 grams** of Wollastonite to the bed to raise the calcium to the desired level. Hopefully this example demonstrates that adding a cup of this or a cup of that really sets you up for overfertilization.

One can question the point of what happens to the nutrient level as the plant grows. Won't the mix become deficient? That would be quite possible if this was the only soluble calcium in the system. Looking back at the standard soil test (Figure 8A), you can see that the value is 4,636 pounds per acre, showing a lot of reserve calcium that should become available as the plants grow. This is an assumption on my part, and running a paste test with available nitrogen in the middle of the growing season would let us know if any amendments are necessary.

Now that the calcium is balanced, we need to rebalance the magnesium and potassium. You could balance each nutrient separately by using magnesium sulfate (Epsom salts) for magnesium and potassium sulfate or sunflower hull ash for potassium. That is perfectly fine, providing you can get the ingredients at a reasonable price and you are someone who prefers to be extremely precise. I prefer to keep the process as simple as possible, within reason. For example, the potassium and magnesium deficiencies could be solved with one product, K-Mag (Sul-Po-Mag). Looking at the analysis of K-Mag, you will see that there is roughly a 1:2 ratio of magnesium to potassium. If we satisfy the magnesium deficiency with K-Mag, we are going to be adding a little extra potassium to the mix. The guidelines that I show in Table 16 are suggestions and not written in stone. There is flexibility with these numbers. It is truly amazing how plants can adjust to their environment if just given a little water and sunlight. Using K-Mag in our calculations to optimize the magnesium in the mix will also fix the potassium at the same time. Looking at Table 9, we see that K-Mag is 11 percent magnesium, and we need to add **9.6 ppm** to our mix, so we can plug those numbers into **Equation 5** to get a result of **0.65 g/ft³**, or a total of **15.6 g** of K-Mag for our 24 ft³ bed:

$$N = \frac{PM}{24C}$$

$$0.65 = \frac{(9.6)(0.18)}{24(0.11)}$$

Since K-Mag has a 2:1 ratio of potassium to magnesium, correcting the 9.6 ppm deficiency of magnesium added 19.2 ppm of potassium to the mix, which is 7.4 ppm more than we needed. For me, the ease of sourcing one product and adding it to the mix justifies a little extra potassium, especially if it doesn't push the soluble salts over the top.

Rebalancing the nitrogen for a peat-corrected mix

Rebalancing the nitrogen starts by going back and seeing how much the original mix had to be diluted in order to get the soluble salts in line with our desired level. In our example, the original mix was diluted with 60 percent Canadian peat, which means the nitrogen is 40 percent of what it was in the original mix. The original mix had 159.0 ppm, but now has only 40 percent of that amount—63.6 ppm. When using something like Canadian peat, I assume the peat contributes zero nitrogen to the mix. The available nitrogen analysis in Figure 9A confirms that assumption, showing only 3.6 ppm of nitrate and ammonium in the peat.

If you were using another bulking source that had a significant amount of nitrogen in it, say 30 ppm, you would multiply that by the percent of bulking agent used to dilute the original mix down (in this case 60 percent, equaling 18 ppm), add that to the nitrogen left in the problem mix, and compare the total to the required level in Table 16. With 63.6 ppm after the Canadian peat dilution, we really have enough nitrogen to grow almost any crop in that table, except long-term cannabis production.

Example 6: After diluting a problem mix with peat, the total available nitrogen in the mix is **23 ppm**. We would like to grow tomatoes, which need **70 ppm**. How much blood meal should we add to each cubic foot of mix to raise the available nitrogen to 70 ppm?

Solution 6: Once again, we will use **Equation 5**. To raise the nitrogen level from 23 ppm to 70 ppm, we need to add **47 ppm** of nitrogen. Media density is not used in the calculation of nitrogen needs like it is in calculating things like calcium, magnesium, and potassium, so we will just use a value of **1** for the media density factor. Table 4 tells us that blood meal is **13 percent** nitrogen. When we plug in these values, we can solve Equation 5 to get a solution of **15.1 g/ft^3**. That means we need to apply 15.1 grams of blood meal for each cubic foot of mix.

$$N = \frac{PM}{24C}$$

$$15.1 = \frac{(47)(1)}{(24)(0.13)}$$

Rebalancing phosphorus in peat-corrected mix

In this example no phosphorus was needed, since the soluble phosphorus after dilution was 3.1 ppm and the phosphorus on the standard test after dilution was 551.2 lb P_2O_5/ac. But what if the phosphorus on the standard test came out to be 150 lb P_2O_5/ac after dilution? What should we do? Even if the soluble phosphorus on the paste test was good, I still would add phosphorus to bring the standard test up to 350–400 lb P_2O_5/ac to make sure that I had plenty of reserve phosphorus. Since phosphorus is always reported in pounds per acre instead of in ppm like the cations, we use Equation 6 to calculate the amount of phosphate product to apply in grams per cubic foot of media. This equation is derived by dividing Equation 5 by 2, because the tests results are given in pounds per acre at a 6-inch, or 0.5-foot, depth.

Equation 6: Amount of phosphate product to apply, in grams per cubic foot

$$N = \frac{PM}{48C}$$

P = Pounds per acre of phosphate to add to the mix
M = Media density, in g/cm³
C = Concentration of P_2O_5 in the amending product, in decimal form
N = Amount of amendment to add, in g/ft³

Or, if we prefer to have the results in pounds per cubic foot, we can use Equation 7. This equation is derived by dividing by the number of cubic feet per acre at a 6-inch depth, which is 21,780 cubic feet.

Equation 7: Amount of phosphate product to apply, in pounds per cubic foot

$$N = \frac{PM}{21{,}780C}$$

Example 7: After diluting a problem mix with peat, it only contains **150 lb P_2O_5/ac**. The media density is **0.25**. How much bone meal, in grams per cubic foot, do we need to add to raise the phosphorus level to **350 pounds P_2O_5 per acre**? How many pounds is that per cubic foot?

Solution 7: This time, we use **Equation 6** because it is specific for phosphate. We need to raise the phosphate level by **200 lb P_2O_5/ac**. Table 7 tells us that bone meal has a phosphate concentration of **22 percent P_2O_5**. And we already know that the media density is **0.25**. Now we can plug these numbers into Equation 6, to get a result of **4.73 g/ft³**:

$$N = \frac{PM}{48C}$$

$$4.73 = \frac{(200)(0.25)}{(48)(0.22)}$$

To get the solution in pounds per cubic foot, we use **Equation 7** instead to get 0.0104 lb bone meal per cubic foot, or **10.4 lb/1,000 ft³**:

$$N = \frac{PM}{21{,}780C}$$

$$0.0104 = \frac{(200)(0.25)}{(21{,}780)(0.22)}$$

Choosing bone meal as the source for correcting a phosphorus deficit will also add calcium, which should be taken into consideration when correcting any calcium deficits in the mix.

When trying to rebalance the mineral side of a mix, list all the cations and anions that need correcting and then choose the amendments that you have available. As in the previous example where we used K-Mag to fix a magnesium deficit and at the same time fixed a potassium shortage, this is a great way to limit the number of purchases and products left over on your shelves. Keep it simple and cost effective. Even if you have a great market for your produce now, it may not always be that way, especially with open-to-market competition.

Rebalancing trace elements on the standard soil report

I have developed a standardized trace element correction package that I use on normal soils, which has worked very well and can be adapted to the modified organic soils using the media density in the calculations. When the trace elements fail to meet my desired levels in normal soils, I use standardized levels for the addition of the different trace elements.

Table 17: Recommended levels of trace elements to correct deficiencies in pounds per 500 cubic feet (1,000 ft^2, 6 inches deep) and grams per cubic foot

Element	Product	% element	500 ft^2	g/ft^3
Boron	Borax	10	0.4	0.36
Manganese	Manganese sulfate	27	0.5	0.45
Copper	Copper sulfate	25	0.2	0.18
Zinc	Zinc sulfate	36	0.5–1.0*	0.45–0.91
*When phosphorus is excessive I use the 1.0 lb rate of zinc sulfate				

These amounts are what I would use on a regular soil, but in growing media I would multiply the above amounts by the media density. For example, in a growing medium with very high phosphorus on the standard soil report, low zinc and boron, and a media density of 0.25, I would apply 0.25 pounds of zinc sulfate and 0.1 lb of Borax per 500 ft^3.

For those who decide that is too complicated or only have a couple of pots and don't want to go into that much detail, I have a default to fall back on: Apply 10 pounds per 1,000ft^2 or 500ft^3 of finely ground Azomite on normal soils, but still multiply by the media density for growing media.

Properly balanced living soil at Sensicare dispensary creates high-quality crops.
(Courtesy of Sensicare)

Conclusion

This is my approach that I have settled on after working with growing media for the last ten years. I am sure there will be some changes down the road. I certainly do not have all the answers but hopefully this information will help you be a better grower in modified media. Please take this information, use it, or reform it and come up with something even better. I have put this information together because I have seen the frustration that growers have trying to grow in media. Best of luck to all of you and may God bless you in your growing endeavors.

Bill McKibben

Properly balanced media-based soil produced this beautiful turmeric and massive "dwarf apple" bananas—all certified organic.
(Courtesy of Stephen Filipiak)

Index

A
aeration, 17, 47
aluminum, 25, 30–31, 34
ammonium, 22–25, 35. *See also* nitrogen
available nitrogen test, 22, 24
Azomite, 45, 56, 95

B
balancing. *See* nutrient balancing
bark fines: analysis of, 14–15; as bulking agent, 1, 10; C:N ratio of, 10, 16; effect on media density, 32
base saturation, 25, 29, 34. *See also* total exchange capacity (TEC)
bicarbonate, 25, 33, 35
biochar, 16, 18, 25
blood meal, 22–24, 52
bone meal, 53, 94
Borax, 45, 55, 94–95. *See also* boron
boron: balancing, 80, 94–95; desired test value for, 25, 30, 34; impact on plant growth, 45–46; leaching of, 60; sources of, 45, 55
brassicas, 79–80
buffering capacity, 37
bulk density, xxi–xxii, 22. *See also* media density
bulking agents: major, 1–10; secondary, 1, 16–18
bulking mix, 63–66, 69–72. *See also* growing media

C
calcium: balancing for, 88–89; for cannabis, 87; desired test value for, 24, 27, 34; impact on plant growth, 42; in irrigation water, 27; Mehlich-3 extraction of, 24; movement of, 8, 59; in sand, 10; sources of, 54, 88
calcium borate, 45
calcium:magnesium (Ca:Mg) ratio, 28
calcium silicate (Wollastonite), 42, 51, 54, 88
calculations: for applying amendments, 73–78; for diluting problem mix, 70–72; for rebalancing, 72–81, 86–95
cannabis, xx, 27, 72, 79–80, 87
cation exchange capacity (CEC). *See* total exchange capacity (TEC)
carbon:nitrogen (C:N) ratio: adjusting, 22–23; of composting materials, 13; desired test value for, 16, 25, 32, 34; impact of raw materials on, 2; measuring, 24
chloride, 25, 33, 35
clay, 30
C:N ratio. *See* carbon:nitrogen (C:N) ratio
cobalt, 25, 32, 34
compost, 1–8, 13, 20–21, 46
containers, 61.
copper: balancing, 80, 94–95; chelation of, 48; desired test value for, 25, 31, 34; impact on plant growth, 48–49; sources of, 55
crop failure, xxi

D
deficiency symptoms, 38–39
dilution, 2, 64, 69–72, 86–95
dolomite. *See* limestone

E
electrical conductivity (EC): converting to soluble salts, 73; desired test value for, 25, 32, 35; excessive, 60; measuring, 32, 69. *See also* soluble salts
feather meal, 22, 24, 52
flowers, 79–80

G
greens, 27, 79–80
Growers Guide for Balancing Soils, 37
grams per cubic foot, conversion to, 72–78
growing media: analysis of, 23–24; balancing nutrients in, xxii, 61, 63–81; crop failures in, xxi; definition of,

xx–xxii; desired test values for, 24–25, 34–35; how to make, 1–18; nitrogen in, 79–80
growing substrate. *See* growing media

H
hydrogen, exchangeable, 25, 29, 34. *See also* pH

I
ideal soil levels, 38
iron: desired test value for, 25, 30, 34; impact of clay on test value, 30; impact on plant growth, 46–47; sources of, 55
irrigation. *See* water, irrigation

K
K-Mag. *See* potassium magnesium sulfate (K-Mag)

L
leaching: 8, 61. *See also* nutrient movement
leaf fines, 1–2, 10, 14–16
limestone, 42–43, 54, 88
living soil. *See* growing media
Logan Labs, xx, xxii, 22

M
magnesium: balancing, 90; for cannabis, 87; desired test value for, 24, 28, 34; impact on plant growth, 43; leaching of, 59; sources of, 54
manganese: balancing, 80, 94–95; desired test value for, 25, 30, 34; impact on plant growth, 47–48; sources of, 55
manure: analysis of, 3–8; in growing media, 2–3, 18; nutrients in, 52–53; and phosphate, 41; and sodium, 28; and soluble salts, 8, 22. *See also* compost
media density: and applying amendments, 61, 95; definition of, xxi–xxii; desired test value for, 25, 32, 35; importance of, xxii, 60, 73, 75; typical, 22.
Mehlich-3 extraction, 24, 26. *See also* soil test, standard
microbial activity, xxi. *See also* soil organisms
mobility of nutrients in the plant, 38
molybdenum: desired test value for, 25, 32, 35; impact on plant growth, 50–51; leaching of, 60; sources of, 56
mono-silicic acid, 51
muck soils, 30

municipal water. *See* water, irrigation

N
nitrate: on available nitrogen test, 22, 24; desired test value for, 25, 35; leaching of, 59. *See also* nitrogen
nitrogen: balancing, 79–80, 91–92; and C:N ratios, 16; for cannabis, 87; impact on plant growth, 39; and molybdenum, 50; sources of, 52; testing for, 63
nutrient balancing: after adding peat, 86–95; desired test values for, 29, 79; in field soils, xx; in growing media, xxii, 61, 69–81; of soluble salts, 64
nutrient deficiencies, xix
nutrient movement, 35. *See also* leaching
nutrients: desired test values for, 24–35; in diluted problem mix, 87; impact on plant growth, 39–51; role in plant, 38; sources of, 52–56

O
organic matter, xxi, 24, 26, 34

P
particle size of amendments, 29
paste test, saturated, xx, 22–24
peat: as bulking agent, 1, 3; density of, 10; for diluting problem mix, 71, 83–95; nitrogen content of, 91; test results for, 9, 84–85
perlite, 17
pH: and bicarbonate, 33–34; of bulking mix, 64; effect on nutrient solubility, 30, 78, 88; and exchangeable hydrogen, 29; and manganese availability, 47; measuring, 24; optimal, 23–24, 26, 34; and sodium, 28
phloem transport of nutrients, 38
phosphorus: balancing, 78, 92–94; for cannabis, 87; desired test value for, 24, 26–27, 34; impact on plant growth, 41; insolubility of, 60; reaction with bicarbonate, 34; sources of, 53; and zinc tie-up, 31
pounds per acre, conversion from, 72
potassium: balancing, 90; for cannabis, 87; desired test value for, 24, 28, 34; impact on plant growth, 44; leaching of, 60; and magnesium, 28, 43; sources of, 55
potassium chloride extraction, 22, 24
potassium magnesium sulfate (K-Mag), 43–44, 53, 55, 90
potassium sulfate, 44
pots. *See* containers

potting mix. *See* growing media
problem mix, 67–72, 86–95. *See also* growing media
pumice, 17

R

rainwater, 57–59. *See also* water, irrigation
rebalancing. *See* nutrient balancing
reverse osmosis (RO) water, 57–59. *See also* water, irrigation
rice hulls, 16, 18
rock phosphate, 27, 53–54. *See also* phosphorus

S

salt accumulation. *See* soluble salts
salt burn, 60. *See also* soluble salts
salt tolerance, 61
sand, 10–12
saturated soils, 8
sea-based products, 28
selenium, 25, 32, 35
seed starter mix, 87
silicon: desired test value for, 25, 32, 35; extraction of, 24; impact on plant growth, 51; in rice hulls, 16; sources of, 56
sodium, 24, 28, 34
soilless media. *See* growing media
soil balancing. *See* nutrient balancing
soil organisms, xx
soil test, standard, 22–24, 34–35, 78
soluble salts: accumulation of, 2, 60; balancing, 64–72; definition of, 32–33; desired test value for, 25, 32, 35, 70; excessive, 26, 35, 60, 64; and germination, 33; impact on TEC, 26; and irrigation water, 59; in manure-based compost, 8, 22; measuring, 32, 60, 69. *See also* electrical conductivity (EC)
stripping waters, 59. *See also* water, irrigation
Styrofoam, 17
sulfur: desired test value, 24, 26, 34; elemental, 40, 53; for cannabis, 87; impact on plant growth, 40; in saturated soils, 8; sources of, 53
sunflower hull ash, 44, 55
surface water, 57–58. *See also* water, irrigation

T

tissue analysis, xx, 24
tomatoes, 27, 79–80
total exchange capacity (TEC), xx, 24–25, 34, 73
toxicity symptoms, 39
trace elements: balancing, 80–81, 94–95; impact on plant growth, 45; sources of, 55
transplants, 32–33, 61, 70

V

vermiculite, 17

W

water, irrigation: precipitating out phosphate, 27; and soluble salts, xxii, 2, 33, 58–59, 61; sources of, 57–58
well water, 57–58. *See also* water, irrigation
Wollastonite. *See* calcium silicate
wood ash, 43, 54–55

X

xylem transport of nutrients, 38

Z

zinc: balancing, 80, 94–95; desired test value for, 25, 31, 34; impact on plant growth, 49–50; sources of, 55–56
zinc sulfate, 49
zeolite, 17–18, 25

Other Titles from Acres U.S.A. Books:

A Growers Guide for Balancing Soils
By William McKibben

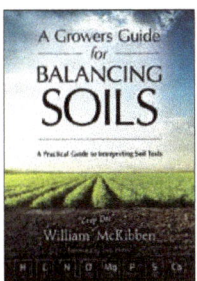

Whether you're a serious home gardener or a full-time farmer, *A Growers Guide for Balancing Soils* will help you expand your understanding of the science behind plant nutrition. It all starts with the soil, but, as you'll read, just spreading compost and hoping for the best isn't a strategy for success. Drawing on 40 years of experience using the Albrecht philosophy of balancing soils, William McKibben will walk you through a data-driven, time-tested process that starts with soil analysis—but doesn't stop there. Productive soil that has the right balance of bio-available minerals and maximizes crop production and quality is the goal, and McKibben outlines a common-sense approach for how to get there.

The Biological Farmer
By Gary Zimmer

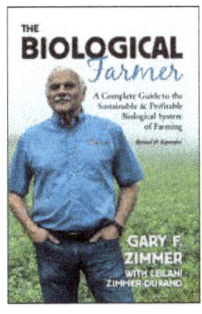

Biological farmers work with nature, feeding soil life, balancing soil minerals, and tilling soils with a purpose. The methods they apply involve a unique system of beliefs, observations, and guidelines that result in increased production and profit. This practical how-to guide explains their methods and will help you make farming profitable and fun. Biological farming does not mean less production; it means eliminating obstacles to healthy, efficient production. Once the chemical, physical, and biological properties of the soil are in balance, you can expect optimal outputs, even in bad years. Biological farming improves the environment, reduces erosion, reduces disease and insect problems, and alters weed pressure—and it accomplishes this by working in harmony with nature.

The Mystery of the Dying Giant Pumpkins
By Grandpa McKibben

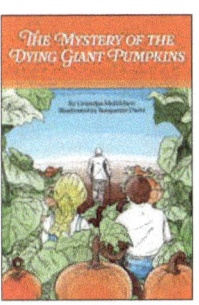

A unique children's book that combines fun and adventure with regenerative agronomy! Becky and Kurt are gearing up for the biggest event of the summer: the Allen County Fair. But when prize-winning pumpkins start mysteriously dying, their grandpa suspects foul play. With a little science, some sneaky detective work, and help from a scarecrow with a secret, the kids join the hunt to uncover the truth. *The Mystery of the Dying Giant Pumpkins* is a fun, fair-filled adventure about gardening, grit, and agronomy.

bookstore.acresusa.com

Visit acresusa.com to learn about our magazine, events, books, and other educational opportunities.